Conceptual
Physical
Science
Explorations

Paul G. Hewitt
John Suchocki
Leslie A. Hewitt

Practice Book

Addison
Wesley

San Francisco Boston New York
Capetown Hong Kong London Madrid Mexico City
Montreal Munich Paris Singapore Sydney Tokyo Toronto

Cover Credit: "Primal Motion"
Michael T. Stewart ©/Molten Images.com

ISBN: 0-321-05184-X

16 17 18 19 20 21 22 - MAL - 10 09 08 07 06

www.aw.com/physics

Welcome to the Conceptual Physical Science

EXPLORATIONS
Practice Book

These practice sheets are explorations to supplement *Conceptual Physical Science—Explorations*. Their purpose is as the name implies—practice—explore—not testing. You should find that exploring and learning physical science is a very central part of your education.

Enjoy!

AFTER completing these pages, look at our answers in the back.

Paul G. Hewitt

John Suchocki

Leslie A. Hewitt

Table of Contents
Conceptual Physical Science Explorations Practice Book

Part 3 Electricity and Magnetism

Part 4 Waves—Sound and Light

Part 5 The Atom

Part 6 Chemistry

Part 7 Earth Science

Chapter 35 Our Natural Landscape

Chapter 36 A Brief History of the Earth

Chapter 37 The Atmosphere, The Oceans, and Their Interactions

Chapter 38 Weather

Part 8 Astronomy

Chapter 39 The Solar System

Chapter 40 The Stars

Appendices

Appendix C Vectors

Appendix D Fluid Physics

Answers

CONCEPTUAL PHYSICAL SCIENCE **EXPLORATIONS**

Chapter 1 About Science
Making Hypotheses

The word science comes from Latin, meaning "to know."
The word *hypothesis* comes from Greek, "under an idea."
A hypothesis (an educated guess) often leads to new
knowledge and may help to establish a theory.

Examples:

1. It is well known that most objects expand when heated. An
   iron plate gets slightly bigger, for example, when put in a
   hot oven. But what of a hole in the middle of the plate?
   Will the hole get bigger or smaller when expansion occurs?
   One friend may say the size of the hole will increase, and
   another says it will decrease.

   a. What is your hypothesis about hole size?
       Is there a test for finding out?

   _____

   _____

   b. There are often several ways to test a hypothesis. For example, you can perform a physical
      experiment and witness the results yourself, or you can use the library to find the reported
      results of other investigators. Which of these two methods do you favor, and why?

   _____

   _____

2. Before the time of the printing press, books were hand-copied by
   scribes, many of whom were monks in monasteries. There is the story
   of the scribe who was frustrated to find a smudge on an important page
   he was copying. The smudge blotted out part of the sentence that
   reported the number of teeth in the head of a donkey. The scribe was
   very upset and didn't know what to do. He consulted with other scribes
   to see if any of their books stated the number of teeth in the head of a
   donkey. After many hours of fruitless searching through the library,
   it was agreed that the best thing to do was to send a messenger by
   donkey to the next monastery and continue the search there.
   What would be your advice?

Making Distinctions

Many people don't seem to see the difference between a thing and the
abuse of the thing. For example, a city council that bans skateboarding
may not distinguish between skateboarding and reckless skateboarding.
A person who advocates that a particular technology be banned may not
distinguish between that technology and the abuses of that technology.
There's a difference between a thing and the abuse of the thing.

On a separate sheet of paper, list other examples where use and abuse are
often not distinguished. Compare your list with others in your class.

CONCEPTUAL PHYSICAL SCIENCE **EXPLORATIONS**
Chapter 2 Newton's First Law—the Law of Inertia
Inertia

(Circle the correct answers.)

1. An astronaut in outer space away from gravitational or frictional forces throws a rock. The rock will

    (gradually slow to a stop)

    (continue moving in a straight line at constant speed)

    The rock's tendency to do this is called

    (inertia)   (weight)   (acceleration)

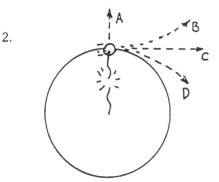

2. The sketch shows a top view of a rock being whirled at the end of a string (clockwise). If the string breaks, the path of the rock is

    (A) (B) (C) (D)

3. Suppose you are standing in the aisle of a bus that travels along a straight road at 100 km/h, and you hold a pencil still above your head. Then relative to the bus, the velocity of the pencil is 0 km/h, and relative to the road, the pencil has a horizontal velocity of

    (less than 100 km/h) (100 km/h) (more than 100 km/h)

    Suppose you release the pencil. While it is dropping, and relative to the road, the pencil still has a horizontal velocity of

    (less than 100 km/h) (100 km/h) (more than 100 km/h)

    This means that the pencil will strike the floor at a place directly

    (behind you)   (at your feet below your hand)   (in front of you)

    Relative to you, the way the pencil drops

    (is the same as if the bus were at rest)

    (depends on the velocity of the bus)

    How does this example illustrate the law of inertia?

    _____

    _____

    _____

CONCEPTUAL PHYSICAL SCIENCE EXPLORATIONS

Chapter 2 Newton's First Law—the Law of Inertia
The Equilibrium Rule: ΣF = 0

1. Little Nellie Newton wishes to be a gymnast and hangs from a variety of positions as shown. Since she is not accelerating, the net force on her is zero. This means the upward pull of the rope(s) equals the downward pull of gravity. She weighs 300 N. Show the scale reading for each case.

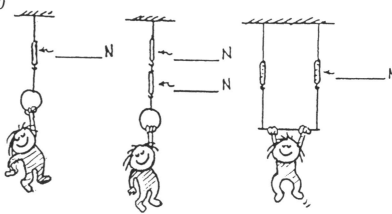

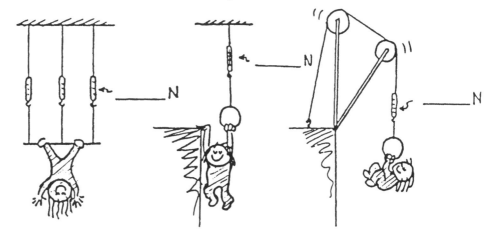

2. When Burl the painter stands in the exact middle of his staging, the left scale reads 600 N. Fill in the reading on the right scale. The total weight of Burl and staging must be

_____N.

3. Burl stands farther from the left. Fill in the reading on the right scale.

4. In a silly mood, Burl dangles from the right end. Fill in the reading on the right scale.

Chapter 2 Newton's First Law—the Law of Inertia
The Equilibrium Rule: ΣF = 0

1. Manuel weighs 1000 N, and stands in the middle of a board that weighs 200 N. The ends of the board rest on bathroom scales. (We can assume the weight of the board acts at its center). Fill in the correct weight reading on each scale.

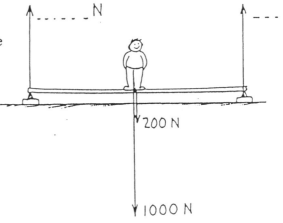

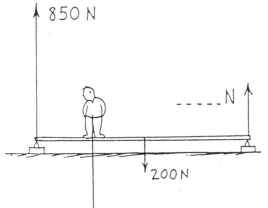

2. When Manuel moves to the left as shown, the scale closest to him reads 850 N. Fill in the weight reading for the far scale.

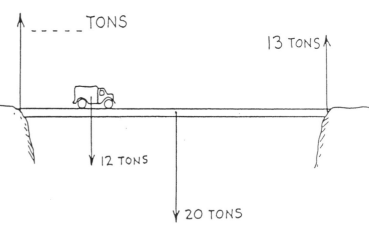

3. A 12-ton truck is one-quarter the way across a bridge that weighs 20 tons. A 13-ton force supports the right side of the bridge as shown. How much support force is on the left side?

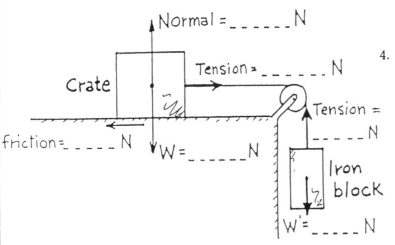

4. A 1000-N crate resting on a horizontal surface is connected to a 500-N iron block through a frictionless pulley as shown. Friction between the crate and surface is enough to keep the system at rest. The arrows show the forces that act on the crate and the block. Fill in the magnitude of each force.

5. If the crate and block in the preceding question move at constant speed, the tension in the rope   (is the same)   (increases)   (**decreases**).

The sliding system is then in   (static equilibrium)   (dynamic equilibrium).

CONCEPTUAL PHYSICAL SCIENCE **EXPLORATIONS**

Chapter 2, 3, or Appendix B: Motion
Free Fall Speed

1. Aunt Minnie gives you $10 per second for 4 seconds.
   How much money do you have after 4 seconds?   _____

2. A ball dropped from rest picks up speed at 10 m/s per second.
   After 4 seconds, how fast is it falling?   _____

3. You have $20, and Uncle Harry gives you $10 each second for 3 seconds.
   How much money do you have after 3 seconds?   _____

4. A ball is thrown straight down with an initial speed of 20 m/s. After 3 seconds,
   how fast is it falling?   _____

5. You have $50 and you pay Aunt Minnie $10/second. When will your money run out?_____

6. You shoot an arrow straight up at 50 m/s. When will it run out of speed?_____

7. So what will be the arrow's speed 5 seconds after you shoot it?   _____

Speed of free fall = acceleration × time = 10 m/s^{2}t.

Average Speed

8. The arrow's initial upward speed is 50 m/s, and 5 seconds later it's zero.
   What's the average speed of the arrow during this time?   _____

9. What is the average speed of the arrow on the way back down—
   that is, for a beginning speed of zero and final speed of 50 m/s?   _____

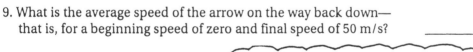

Average speed = $\dfrac{\text{initial speed} + \text{final speed}}{2}$

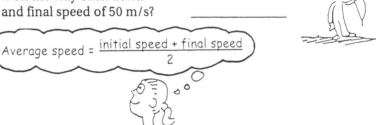

Free Fall Distance

10. Speed is one thing; distance another. How high will the arrow be 5 seconds after being shot up at
    50 m/s?   _____

11. If an apple falls from a tree and hits the ground below in 1 second, its speed is 10 m/s when it hits.
    In words, explain why it falls a vertical distance of 5 meters, and NOT 10 meters.

_____   Distance = average speed × time.

_____

_____

CONCEPTUAL PHYSICAL SCIENCE **EXPLORATIONS**
Chapter 3 Newton's Second Law—Force and Acceleration

Free Fall

A rock dropped from the top of a cliff picks up speed as it falls. Pretend that a speedometer and odometer are attached to the rock to show readings of speed and distance at 1-second intervals. Both speed and distance are zero at time = zero (see sketch). Note that after falling 1 second the speed reading is 10 m/s and the distance fallen is 5 m. The readings for succeeding seconds of fall are not shown and are left for you to complete. So draw the position of the speedometer pointer and write in the correct odometer reading for each time. Use g = 10 m/s² and neglect air resistance.

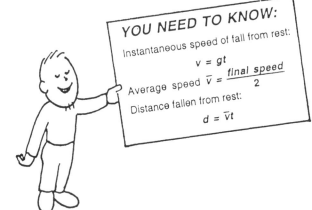

YOU NEED TO KNOW:

Instantaneous speed of fall from rest:
$$v = gt$$

Average speed $\bar{v} = \dfrac{final\ speed}{2}$

Distance fallen from rest:
$$d = \bar{v}t$$

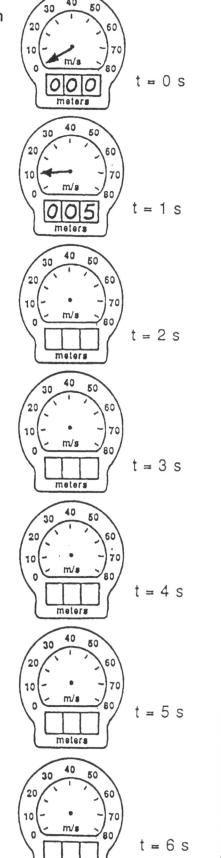

t = 0 s

t = 1 s

t = 2 s

t = 3 s

t = 4 s

t = 5 s

t = 6 s

1. The speedometer reading increases by the same amount, _____ m/s, each second. This increase in speed per second is called _____.

2. The distance fallen increases as the square of the _____.

3. If it takes 7 seconds to reach the ground, then its speed at impact is _____ m/s, the total distance fallen is _____ m, and its acceleration of fall just before impact is _____ m/s².

Chapter 3, or Appendix B: (Motion)
Hang Time

Some athletes and dancers have great jumping ability. When they leap straight up, they seem to momentarily "hang in the air" and defy gravity. The time that a jumper is airborne with feet off the ground is called **hang time**. Ask your friends to estimate the hang time of the great jumpers. They may say two or three seconds. But surprisingly, the hang time of the greatest jumpers is most always less than 1 second! A longer time is one of many illusions we have about nature.

To better understand this, find the answers to the following questions:

1. If you step off a table and it takes one-half second to reach the floor, what will be your speed when you meet the floor?

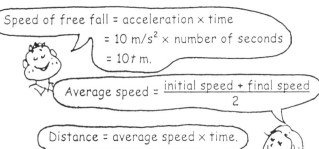

Speed of free fall = acceleration × time
= 10 m/s² × number of seconds
= 10t m.

2. What will be your average speed of fall?

Average speed = $\frac{\text{initial speed} + \text{final speed}}{2}$

3. What will be the distance of fall?

Distance = average speed × time.

4. So how high is the surface of the table above the floor?

Jumping ability is best measured by a standing vertical jump. Stand facing a wall with feet flat on the floor and arms extended upward. Make a mark on the wall at the top of your reach. Then make your jump and at the peak make another mark. The distance between these two marks measures your vertical leap. If it's more than 0.6 meters (2 feet), you're exceptional.

5. What is your vertical jumping distance?

6. Calculate your personal hang time. (Use the formula d = 1/2 gt², from Appendix B. Remember that hang time is the time that you move upward + the time you return downward.)

Almost anybody can safely step off a 1.25-m (4-feet) high table. Can anybody in your school jump from the floor up onto the same table?

No way!

There's a big difference in how high you can reach and how high you raise your "center of gravity" when you jump. Even basketball star Michael Jordan in his prime couldn't quite raise his body 1.25 meters high, although he could easily reach higher than the more-than-3-meter high basket.

Here we're talking about vertical motion. How about running jumps? We'll see in Chapter 8 that the height of a jump depends only on the jumper's vertical speed at launch. While airborne, the jumper's horizontal speed remains constant while the vertical speed undergoes acceleration due to gravity. While airborne, no amount of leg or arm pumping or other bodily motions can change your hang time.

Fascinating physical science!

CONCEPTUAL PHYSICAL SCIENCE EXPLORATIONS

Chapter 3 Newton's Second Law—Force and Acceleration
Friction Force and Acceleration

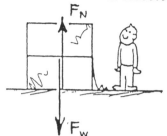

Here's the crate filled with delicious peaches discussed in Chapter 4. The only vertical forces acting on the crate are gravity and the support force of the floor. The vector F_w is the weight, and F_N is the support force. We use the subscript $_N$ to indicate the force is **normal** (right angles) to the floor.

1. When Harry pulls with force **P**, 150 N, and the crate doesn't move, what is the magnitude of the friction f? _____

2. If the pulling force **P** is increased to 200 N and the crate still doesn't move, what is the magnitude of f? _____

3. So when Harry pulls with 200 N, the crate is in
   (static equilibrium)   (dynamic equilibrium).

4. Harry increases his pull and the crate begins to slide. When he pulls with 250 N it slides at constant velocity. What is the magnitude of f? _____

5. When Harry pulls with 250 N the crate is in
   (static equilibrium)   (dynamic equilibrium).

6. If the mass of the crate is 50 kg and sliding friction is 250 N, what is the acceleration of the crate when pulled with 250 N? _____ 300 N? _____ 500 N? _____

Force and Acceleration

1. Skelly the skater, total mass 25 kg, is propelled by rocket power.

   a. Complete Table I
      (neglect resistance)

TABLE I	FORCE	ACCELERATION
	100 N	
	200 N	
		10 m/s²

   b. Complete Table II for a
      constant 50-N resistance.

TABLE II	FORCE	ACCELERATION
	50 N	0 m/s²
	100 N	
	200 N	

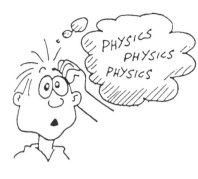

Chapter 3 Newton's Second Law—Force and Acceleration
Mass and Weight

1. Use the words *mass*, *weight*, and *volume*, to complete the table.

The force due to gravity on an object	
The quantity of matter in an object	
The amount of space an object occupies	

2. Different masses are hung on a spring scale calibrated in newtons.

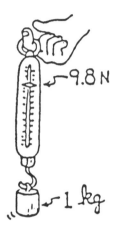

The force exerted by gravity on 1 kg = 9.8 N.

The force exerted by gravity on 5 kg = _____ N.

The force exerted by gravity on _____ kg = 98 N.

Make up your own mass and show the corresponding weight:

The force exerted by gravity on _____ kg = _____ N.

3. By whatever means (spring scales, measuring balance, etc.), find the mass of your physical science book. Then complete Table I.

Table I

OBJECT	MASS	WEIGHT
MELON	1 kg	
APPLE		1 N
BOOK		
UNCLE HARRY	90 kg	

4. Why isn't the girl hurt when the nail is driven into the block of wood?

_____

_____

Would this be more dangerous or less dangerous if the block were less massive _____? Explain.

_____

_____

_____

CAUTION: Safety dictates you not try this experiment yourself.

CONCEPTUAL PHYSICAL SCIENCE **EXPLORATIONS**

Chapter 4 Newton's Third Law—Action and Reaction

Action and Reaction Pairs

1. In the example below, the action-reaction pair is shown by the arrows (vectors), and the action-reaction described in words. In (*a*) through (*g*) draw the other arrow (vector) and state the reaction to the given action. Then make up your own example in (*h*).

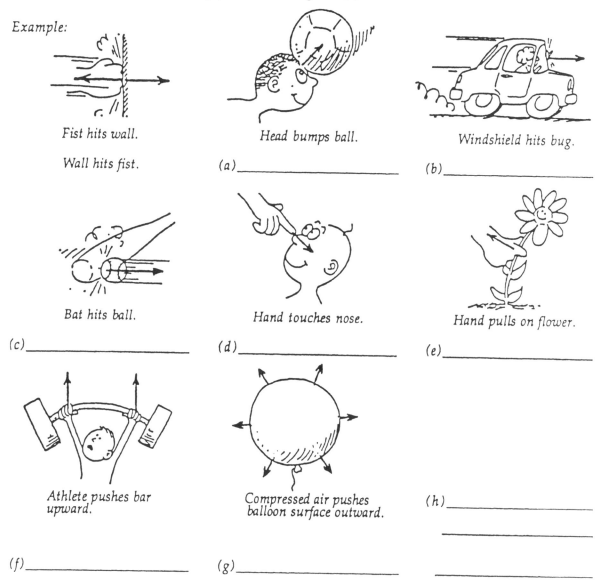

Example:

Fist hits wall.

Wall hits fist.

Head bumps ball.

(*a*)_____

Windshield hits bug.

(*b*)_____

Bat hits ball.

(*c*)_____

Hand touches nose.

(*d*)_____

Hand pulls on *flower.*

(*e*)_____

Athlete pushes bar upward.

(*f*)_____
_____

Compressed air pushes balloon surface outward.

(*g*)_____
_____

(*h*)_____
_____
_____

2. Draw arrows to show the chain of at least six pairs of action-reaction forces below.

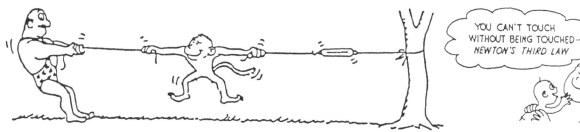

YOU CAN'T TOUCH WITHOUT BEING TOUCHED— NEWTON'S THIRD LAW

CONCEPTUAL PHYSICAL SCIENCE **EXPLORATIONS**

Chapter 5 Momentum
Changing Momentum

1. A moving car has momentum. If it moves twice as fast, its momentum is _____ as much.

2. Two cars, one twice as heavy as the other, move down a hill at the same speed. Compared with the lighter car, the momentum of the heavier car is _____ as much.

3. The recoil momentum of a cannon that kicks is
   (more than) (less than) (the same as)
   the momentum of the cannonball it fires.

   (Here we neglect the momentum of the gases.)

4. Suppose you are traveling in a bus at highway speed on a nice summer day and the momentum of an unlucky bug is suddenly changed as it splatters onto the front window.

   a. Compared to the force that acts on the bug, how much force acts on the bus?
      (more) (the same) (less)

   b. The time of impact is the same for both the bug and the bus. Compared to the impulse on the bug, this means the impulse on the bus is
      (more) (the same) (less)

   c. Although the momentum of the bus is very large compared to the momentum of the bug, the change in momentum of the bus, compared to the change of momentum of the bug is
      (more) (the same) (less)

   d. Which undergoes the greater acceleration?
      (bus) (both the same) (bug)

   e. Which therefore, suffers the greater damage?
      (bus) (both the same) (the bug of course!)

Chapter 5 Momentum

5. Granny whizzes around the rink and is suddenly confronted with Ambrose at rest directly in her path. Rather than knock him over, she picks him up and continues in motion without "braking."

Consider both Granny and Ambrose as two parts of one system. Since no outside forces act on the system, the momentum of the system before collision equals the momentum of the system after collision.

a. Complete the before-collision data in the table below.

BEFORE COLLISION	
Granny's mass	80 kg
Granny's speed	3 m/s
Granny's momentum	_____
Ambrose's mass	40 kg
Ambrose's speed	0 m/s
Ambrose's momentum	_____
Total momentum	_____

b. After collision, does Granny's speed increase or decrease?

   _____

c. After collision, does Ambrose's speed increase or decrease?

   _____

d. After collision, what is the total mass of Granny + Ambrose?

   _____

e. After collision, what is the total momentum of Granny + Ambrose?

   _____

f. Use the conservation of momentum law to find the speed of Granny and Ambrose together after collision.
   (Show your work in the space below.)

New speed = _____

CONCEPTUAL PHYSICAL SCIENCE EXPLORATIONS

Chapter 5 Momentum

Systems

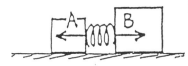

1. When the compressed spring is released, Blocks A and B will slide apart. There are 3 systems to consider here, indicated by the closed dashed lines below — System A, System B, and System A+B. Ignore the vertical forces of gravity and the support force of the table.

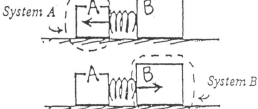

a. Does an external force act on System A?   (yes) (no)   *System A*

 Will the momentum of System A change? (yes) (no)

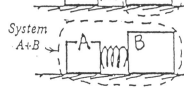

b. Does an external force act on System B?   (yes) (no)   *System B*

 Will the momentum of System B change? (yes) (no)

c. Does an external force act on System A+B?   (yes) (no)   *System A+B*

 Will the momentum of System A+B change? (yes) (no)

Note that external forces on System A and System B are internal to System A+B, so they cancel!

2. Billiard ball A collides with billiard ball B at rest. Isolate each system with a closed dashed line. Draw only the external force vectors that act on each system.

System A   *System B*   *System A+B*

a. Upon collision, the momentum of System A   (increases) (decreases) (remains unchanged).

b. Upon collision, the momentum of System B   (increases) (decreases) (remains unchanged).

c. Upon collision, the momentum of System A+B (increases) (decreases) (remains unchanged).

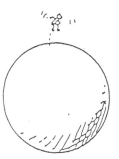

3. A girl jumps upward. In the sketch to the left, draw a closed dashed line to indicate the system of the girl.

a. Is there an external force acting on her? (yes) (no)

 Does her momentum change? (yes) (no)

 Is the girl's momentum conserved? (yes) (no)

b. In the sketch to the right, draw a closed dashed line to indicate the system [girl + earth]. Is there an external force due to the interaction between the girl and the earth that acts on the system? (yes) (no)

 Is the momentum of the system conserved? (yes) (no)

4. A block strikes a blob of jelly. Isolate 3 systems with a closed dashed line and show the external force on each. In which system is momentum conserved?

5. A truck crashes into a wall. Isolate 3 systems with a closed dashed line and show the the external force on each. In which system is momentum conserved?

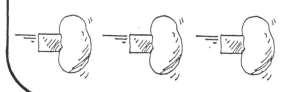     

CONCEPTUAL PHYSICAL SCIENCE EXPLORATIONS

Chapter 6 Work and Energy

Work, Power, and Energy

1. How much work (energy) is needed to lift an object that weighs 200 N to a height of 4 m?

    _____

2. How much power is needed to lift the 200-N object to a height of 4 m in 4 s?

    _____

3. What is the power output of an engine that does 60,000 J of work in 10 s?

    _____

4. The block of ice weighs 500 newtons.

    a. Neglecting friction, how much force is needed to push it up the incline?

        _____

    b. How much work is required to push it up
        3 meters?

        _____

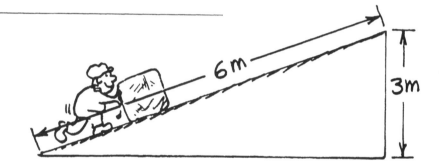

5. All the ramps are 5 m high. We know that the KE of the block at the bottom of the ramp will be equal to the loss of PE (conservation of energy). Find the speed of the block at ground level in **each case.** [Hint: Do you recall from earlier chapters how long it takes something to fall a vertical distance of 5 m from a positon of rest (assume g = 10 m/s²)? And how much speed a falling object acquires in this time? This gives you the answer to Case 1. Cases 2 and 3 look more complex. But not so if you use energy conservation to guide your answer (no further computations needed!). Discuss with your classmates how energy conservation gives you the answers to Cases 2 and 3.]

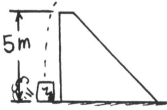

Case 1: Speed = _____ m/s   Case 2: Speed = _____ m/s   Case 3: Speed = _____ m/s.

6. Which block gets to the bottom of the incline first? Assume no friction. (Be careful!) Explain your answer.

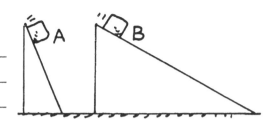

_____

_____

_____

7. The KE and PE of a block freely sliding down a ramp are shown in only one place in the sketch. Fill in the missing values.

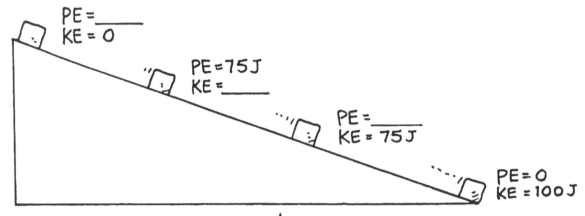

PE = _____
KE = 0

PE = 75 J
KE = _____

PE = _____
KE = 75 J

PE = 0
KE = 100 J

8. A big metal bead slides due to gravity along an upright friction-free wire. It starts from rest at the top of the wire as shown in the sketch. How fast is it traveling as it passes

Point B?_____

Point D?_____

Point E?_____

At what point does it have the

maximum speed?_____

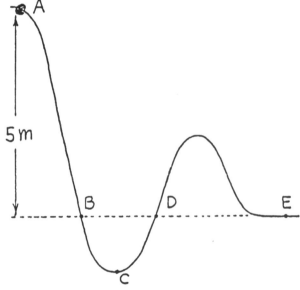

5 m

A
B   D   E
C

9. Rows of wind-powered generators are used in various windy locations to generate electric power. Does the power generated affect the speed of the wind? Would locations behind the 'windmills' be windier if they weren't there? Discuss this in terms of energy conservation with your classmates.

_____

_____

_____

_____

CONCEPTUAL PHYSICAL SCIENCE **EXPLORATIONS**

Chapter 6 Work and Energy
Conservation of Energy

Fill in the blanks for the six systems shown.

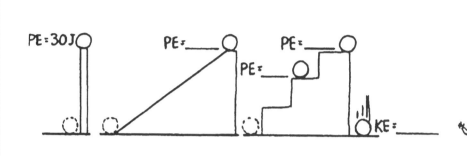

PE = 30 J   PE = ____   PE = ____

PE = ____

KE = ____

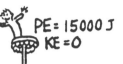

PE = 15000 J
KE = 0

PE = 11250 J
KE = ____

PE = 7500 J
KE = ____

PE = 3750 J
KE = ____

 $v = 30 \frac{km}{h}$
KE = 10^6 J

 $v = 60 \frac{km}{h}$
KE = ____

$v = 90 \frac{km}{h}$
KE = ____

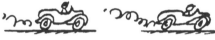

 PE = 0 J
KE = ____

 PE = 10^4 J

WORK DONE = ____

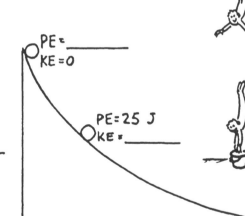

PE = ____
KE = 0

PE = 25 J
KE = ____

PE = 0
KE = 50 J

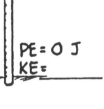

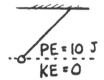 PE = 10 J
KE = 0

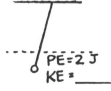 PE = 2 J
KE = ____

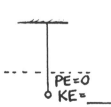 PE = 0
KE = ____

 PE = ____
KE = ____

Chapter 6 Work and Energy
Machines

1. The woman supports a 100-N load with the friction-free pulley systems shown below. Fill in the spring-scale readings that show how much force she must exert.

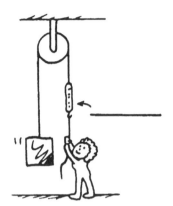  

2. A 600-N block is lifted by the friction-free pulley system shown.

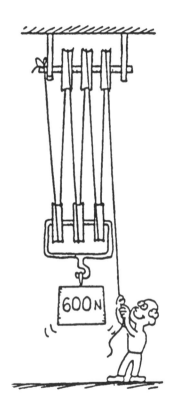

   a. How many strands of rope support the 600-N weight?

   _____

   b. What is the tension in each strand?

   _____

   c. What is the tension in the end held by the man?

   _____

   d. If the man pulls his end down 60 cm, how many cm will the weight rise?

   _____

   e. Does the man multiply force or energy (or both)?

   _____

   f. If the man does 60 joules of work, what will be the increase of PE of the 600-N weight?

   _____

3. Why don't balls bounce as high during the second bounce as they do in the first?

   _____
   _____
   _____

CONCEPTUAL PHYSICAL SCIENCE **EXPLORATIONS**

Chapter 7 Gravity
Inverse–Square Law

1. Paint spray travels radially away from the nozzle of the can in straight lines. Like gravity, the strength (intensity) of the spray obeys an inverse-square law. Complete the diagram by filling in the blank spaces.

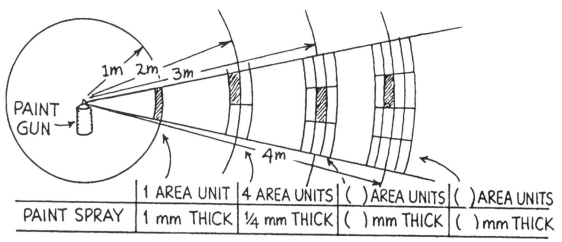

PAINT SPRAY	1 AREA UNIT	4 AREA UNITS	( ) AREA UNITS	( ) AREA UNITS
	1 mm THICK	¼ mm THICK	( ) mm THICK	( ) mm THICK

2. A small light source located 1 m in front of an opening of area 1 m² illuminates a wall behind. If the wall is 1 m behind the opening (2 m from the light source), the illuminated area covers 4 m². How many square meters will be illuminated if the wall is

   5 m from the source?_____

   10 m from the source? _____

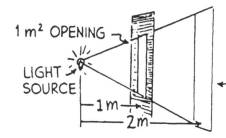

3. The sound of a bullfrog attracts a potential mate. How much quieter is the sound if the potential mate is 5 times farther away?

   _____

4. The sound of a cricket gets stronger as you get closer to it. If you get twice as close, how much louder will the sound be?

   _____

5. You are on a spaceship approaching the planet Jupiter. You feel its gravitational pull. When you get three times closer to Jupiter, how much stronger will be the pull?

   _____

6. At a certain distance from each other, two asteroids in space feel a gravitational attraction to each other of 1000 N. When their distance apart is reduced to half the distance, what will be the gravitational force on each asteroid?

   _____

Chapter 7 Gravity
Force and Weight

1. An apple that has a mass of 0.1 kilogram has the same mass wherever it is. The amount of matter that makes up the apple

    (depends upon)   (does not depend upon)

    the location of the apple. It has the same resistance to acceleration wherever it is — its inertia everywhere is

    (the same)   (different).

    The weight of the apple is a different story. It may weigh exactly 1 N in San Francisco and slightly less in mile-high Denver, Colorado. On the surface of the moon the apple would weigh 1/6 N, and far out in outer space it may have almost no weight at all. The quantity that doesn't change with location is

    (mass)   (weight),

    and the quantity that may change with location is its

    (mass)   (weight).

    That's because

    (mass)   (weight)

    is the force due to gravity on a body, and this force varies with distance. So weight is the force of gravity between two bodies, usually some small object in contact with the Earth.   When we refer to the

    (mass)   (weight)

    of an object we are usually speaking of the gravitational force that attracts it to the Earth.

    Fill in the blanks:

2. If we stand on a weighing scale and find that we are pulled toward the Earth with a force of 500 N,

    then we weigh _____ N. Strictly speaking, we weigh _____ N relative to the Earth. How much does the Earth weigh? If we tip the scale upside down and repeat the weighing process, we can say that we and the Earth are still pulled together with a force of _____ N, and therefore, relative to us, the whole 6,000,000,000,000,000,000,000,000-kg Earth weighs _____ N! Weight, unlike mass, is a relative quantity.

VIEW THE SAME FROM ANOTHER PERSPECTIVE!

DO YOU SEE WHY IT MAKES SENSE TO DISCUSS THE EARTH'S MASS, BUT NOT ITS WEIGHT?

We are pulled to the Earth with a force of 500 N, so we weigh 500 N.

The Earth is pulled toward us with a force of 500 N, so it weighs 500 N.

CONCEPTUAL PHYSICAL SCIENCE **EXPLORATIONS**
   Chapter 7 Gravity
   Our Ocean Tides

1. Consider two equal-mass blobs of water,
   A and B, initially at rest in the moon's
   gravitational field. The vector shows the
   gravitational force of the moon on A.

   a. Draw a force vector on B due to the moon's gravity.

   b. Is the force on B more or less than the force on A? _____

   c. Why?_____

   d. The blobs accelerate toward the moon. Which has the greater acceleration? (A) (B)

   e. Because of the different accelerations, with time

      (A gets farther ahead of B) (A and B gain identical speeds) and the distance between A and B

      (increases) (stays the same) (decreases).

   f. If A and B were connected by a rubber band, with time the rubber band would

      (stretch) (not stretch).

   g. This (stretching) (nonstretching) is due to the (difference) (nondifference) in the moon's
      gravitational pulls.

   h. The two blobs will eventually crash into the moon. To orbit around the moon instead of
      crashing into it, the blobs should move

      (away from the moon) (tangentially). Then their accelerations will consist of changes in

      (speed) (direction).

2. Now consider the same two blobs
   located on opposite sides of the Earth.

   a. Because of differences in the moon's pull on the blobs, they tend to

      (spread away from each other) (approach each other).

   b. Does this spreading produce ocean tides? (Yes) (No)

   c. If Earth and moon were closer, gravitational force between them would be

      (more) (the same) (less), and the difference in gravitational forces on the near and far parts

      of the ocean would be (more) (the same) (less).

   d. Because the Earth's orbit about the sun is slightly elliptical, Earth and sun are closer in December

      than in June. Taking the sun's tidal force into account, on a world average, ocean tides are greater in

      (December) (June) (no difference).

CONCEPTUAL PHYSICAL SCIENCE EXPLORATIONS

Chapter 8 Projectile and Satellite Motion
Projectiles

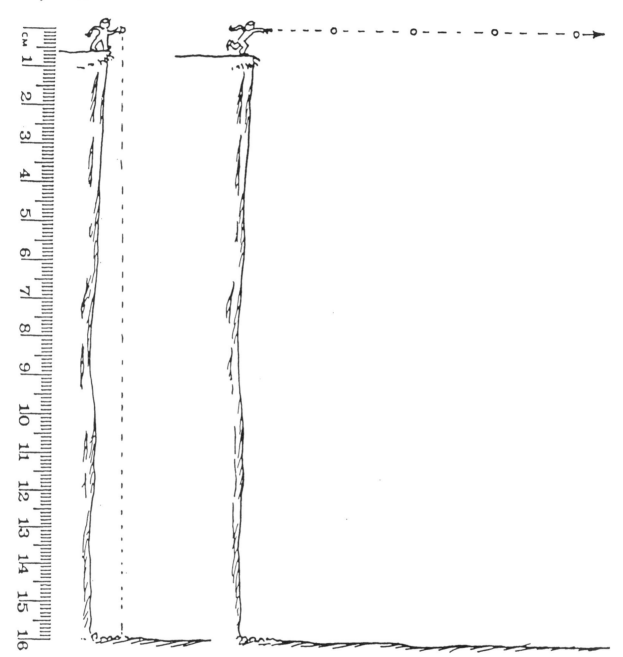

1. Above left: Use the scale 1 cm: 5 m and draw the positions of the dropped ball at 1-second intervals. Neglect air drag and assume $g = 10$ m/s^2. Estimate the number of seconds the ball is in the air.

    _____ seconds.

2. Above right: The four positions of the thrown ball with *no gravity* are at 1-second intervals. At 1 cm: 5 m, carefully draw the positions of the ball *with* gravity. Neglect air drag and assume $g = 10$ m/s^2. Connect your positions with a smooth curve to show the path of the ball. How is the motion in the vertical direction affected by motion in the horizontal direction?

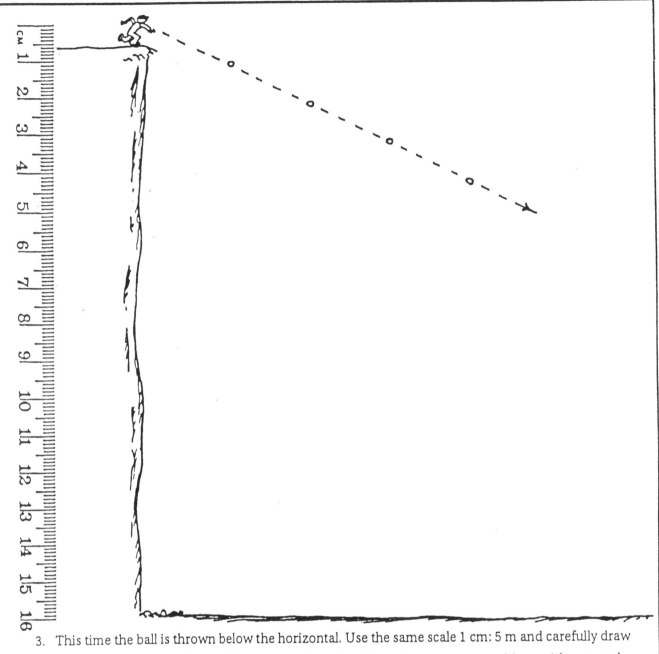

3. This time the ball is thrown below the horizontal. Use the same scale 1 cm: 5 m and carefully draw the positions of the ball as it falls beneath the dashed line. Connect your positions with a smooth curve. Estimate the number of seconds the ball remains in the air._____ s

4. Suppose that you are an accident investigator and you are asked to figure whether or not the car was speeding before it crashed through the rail of the bridge and into the mudbank as shown. The speed limit on the bridge is 55 mph = 24 m/s. What is your conclusion?

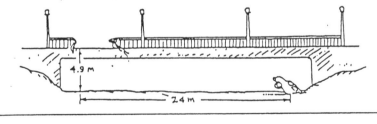

4.9 m

24 m

28

CONCEPTUAL PHYSICAL SCIENCE EXPLORATIONS

Chapter 8 Projectile and Satellite Motion
Circular Orbit

1. Figure A shows "Newton's Mountain," so high that its top is above the drag of the atmosphere. The cannonball is fired and hits the ground as shown.

   Figure A

   a. You draw the path the cannonball might take if it were fired a little bit faster.

   b. Repeat for a still greater speed, but still less than 8 km/s.

   c. Then draw the orbital path it would take if its speed were 8 km/s.

   d. What is the shape of the 8 km/s curve?

      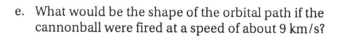

   e. What would be the shape of the orbital path if the cannonball were fired at a speed of about 9 km/s?

      _____

2. Figure B shows a satellite in circular orbit.

   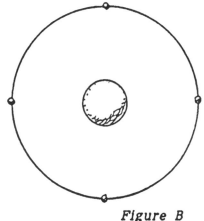
   Figure B

   a. At each of the four positions draw a vector that represents the gravitational **force** exerted on the satellite.

   b. Label the force vectors *F*.

   c. Then draw at each position a vector to represent the **velocity** of the satellite at that position, and label it *V*.

   d. Are all four *F* vectors the same length? Why or why not?

      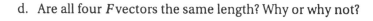

   e. Are all four *V* vectors the same length? Why or why not?

      _____

   f. What is the angle between your *F* and *V* vectors? _____

   g. If an enormously tall bowling alley extended as high as the satellite orbits, can you see that the force of gravity wouldn't change the speed of the ball (because it's always pulling at right angles to the alley)? So what does this tell you about the work that the force of gravity does on a satellite in circular orbit?

      _____

   h. Does the kinetic energy of the satellite in Figure B remain constant, or does it vary? _____

   i. Does the potential energy of the satellite remain constant, or does it vary? _____

3. Figure C shows a satellite in elliptical orbit.

   a. Repeat the procedure you used for the circular orbit, drawing vectors *F* and *V* for each position, including proper labeling. Show equal magnitudes with equal lengths, and greater magnitudes with greater lengths, but don't bother making the scale accurate.

   b. Are your vectors *F* all the same magnitude?
      Why or why not?

   _____

   _____

   c. Are your vectors *V* all the same magnitude?
      Why or why not?

   _____

   _____

   _____

   _____

   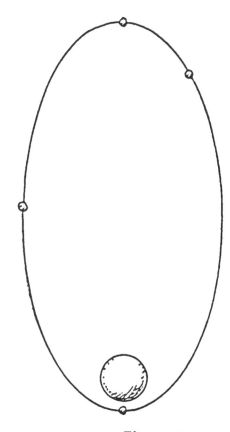

   d. Is the angle between vectors *F* and *V* everywhere the same, or does it vary?

   _____

   e. Are there places where there is a component of *F* along *V*? (That is, are there places where the force of gravity is not at right angles to the orbit.)

   _____

   _____

   f. Is work done on the satellite when there is a component of *F* along and in the same direction of *V* and if so, does this increase or decrease the KE of the satellite?

   _____

   Figure C

   g. When there is a component of *F* along and opposite to the direction of *V*, does this increase or decrease the KE of the satellite?

   _____

   _____

   _____

   h. What can you say about the sum KE + PE along the orbit?

   _____

CONCEPTUAL PHYSICAL SCIENCE EXPLORATIONS
Chapter 9 Thermal Energy
Temperature, Heat, and Expansion

1. Complete the table:

TEMPERATURE OF MELTING ICE	°C	32°F	K
TEMPERATURE OF BOILING WATER	°C	212°F	K

2. Suppose you apply a flame and heat one liter of water, raising its temperature 10°C. If you transfer the same heat energy to two liters, how much will the temperature rise? For three liters? *Record your answers on the blanks in the drawing at the right.*

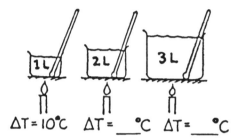

$\Delta T = 10°C$ $\Delta T = \_\_°C$ $\Delta T = \_\_°C$

3. A thermometer is in a container half-filled with 20°C water.

   a. When an equal volume of 20°C water is added, the temperature of the mixture is

   (10°C) (20°C) (40°C)

   b. When instead an equal volume of 40°C water is added, the temperature of the mixture will be

   (20°C) (30°C) (40°C)

   c. When instead a small amount of 40°C water is added, the temperature of the mixture will be

   (20°C) (between 20°C and 30°C) (30°C) (more than 30°C)

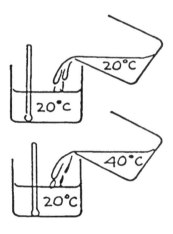

4. A red-hot piece of iron is put into a bucket of cool water. *Mark the following statements true (T) or false (F).* (Ignore heat transfer to the bucket.)

   a. The decrease in iron temperature equals the increase in the water temperature._____

   b. The quantity of heat lost by the iron equals the quantity of heat gained by the water. _____

   c. The iron and water both will reach the same temperature. _____

   d. The final temperature of the iron and water is halfway between the initial temperatures of each. _____

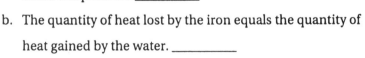

CAN COMMON ICE BE COLDER THAN 0°C?

Chapter 9 Thermal Energy
Thermal Expansion

1. The weight hangs above the floor from the copper wire. When a candle is moved along the wire and heats it, what happens to the height of the weight above the floor? Why?

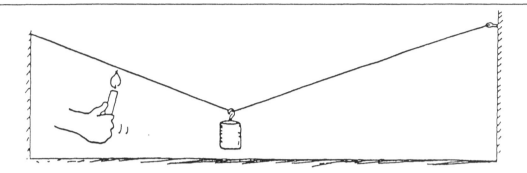

2. The levels of water at 0°C and 1°C are shown below in the first two flasks. At these temperatures there is microscopic slush in the water. There is slightly more slush at 0°C than at 1°C. As the water is heated, some of the slush collapses as it melts, and the level of the water falls in the tube. That's why the level of water is slightly lower in the 1°C-tube. Make rough estimates and sketch in the appropriate levels of water at the other temperatures shown. What is important about the level when the water reaches 4°C?

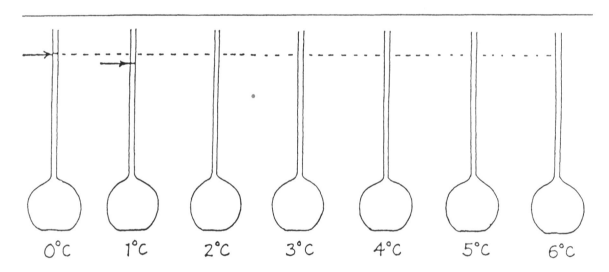

0°C   1°C   2°C   3°C   4°C   5°C   6°C

3. The diagram at right shows an ice-covered pond. Mark the probable temperatures of water at the top and bottom of the pond.

32

CONCEPTUAL PHYSICAL SCIENCE EXPLORATIONS

Chapter 10 Heat Transfer and Change of Phase
Evaporation

1. Why does it feel colder when you swim at a pool on a windy day?
   _____

2. Why does your skin feel cold when a little rubbing alcohol is applied to it?
   _____

3. Briefly explain from a molecular point of view why evaporation is a cooling process.
   _____
   _____

4. When hot water rapidly evaporates, the result can be dramatic. Consider 4 g of boiling water spread over a large surface so that 1 g rapidly evaporates. Suppose further that the surface and surroundings are very cold so that all 540 calories for evaporation come from the remaining 3 g of water.

   a. How many calories are taken from each gram of water?
   _____

   b. How many calories are released when 1 g of 100°C water cools to 0°C?
   _____

   c. How many calories are released when 1 g of 0°C water changes to 0°C ice?
   _____

   d. What happens in this case to the remaining 3 g of boiling water when 1 g rapidly evaporates?
   _____
   _____

Energy is nature's way of keeping score!

CONCEPTUAL PHYSICAL SCIENCE EXPLORATIONS

Chapter 10 Heat Transfer and Change of Phase
Change of Phase

All matter can exist in the solid, liquid, or gaseous phases. The solid phase exists at relatively low temperatures, the liquid phase at higher temperatures, and the gaseous phase at still higher temperatures. Water is the most common example, not only because of its abundance but also because the temperatures for all three phases are common.

1. How many calories are needed to change 1 gram of 0°C ice to water?

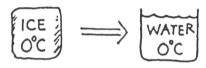

_____

2. How many calories are needed to change the temperature of 1 gram of water by 1°C?

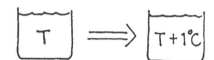

_____

3. How many calories are needed to melt 1 gram of 0°C ice and turn it to water at a room temperature of 23°C?

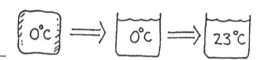

_____

4. A 50-gram sample of ice at 0°C is placed in a glass beaker that contains 200 g of water at 20°C.

    a. How much heat is needed to melt the ice? _____

    b. By how much would the temperature of the water change if it gave up this much heat to the ice? _____

    c. What will be the final temperature of the mixture? (Disregard any heat absorbed by the glass or given off by the surrounding air.) _____

5. How many calories are needed to change 1 gram of 100°C boiling water to 100°C steam?

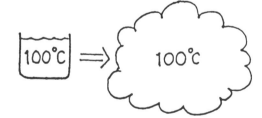

_____

6. Fill in the number of calories at each step below for changing the state of 1 gram of 0°C ice to 100°C steam.

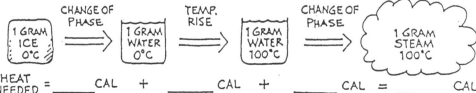

Chapter 11 Electricity

Coulomb's Law

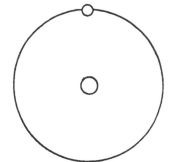

1. The diagram is of a hydrogen atom.

   a. Label the proton in the nucleus with a + sign and the orbital electron with a - sign.

   b. The electrical interaction between the nucleus and the orbital electron is a force of

   (attraction) (repulsion)

   c. According to Coulomb's law,

   $$F = k \frac{q_1 q_2}{d^2}$$

   if the charge of either the nucleus or the orbital electron were greater, the force between the nucleus and the electron would be

   (greater) (less)

   and if the distance between the nucleus and electron were greater the force would be

   (greater) (less).

   If the distance between the nucleus and electron were doubled, the force would be

   (1/4 as much) (1/2 as much) (two times as much) (4 times as much)

2. Consider the electric force between a pair of charged particles a certain distance apart. By Coulomb's law:

   a. If the charge on one of the particles is doubled, the force is
      (unchanged) (halved) (doubled) (quadrupled)

   b. If, instead, the charge on both particles is doubled, the force is
      (unchanged) (halved) (doubled) (quadrupled)

   c. If instead the distance between the particles is halved, the force is
      (unchanged) (halved) (doubled) (quadrupled)

   d. If the distance is halved, and the charge of both particles is doubled,

      the force is _____ times as great.

CONCEPTUAL PHYSICAL SCIENCE **EXPLORATIONS**

Chapter 11 Electricity
Electric Pressure—Voltage

1. 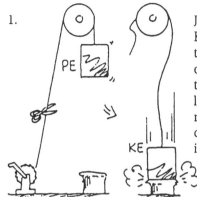 Just as PE (potential energy) transforms to KE (kinetic energy) for a mass lifted against the gravitational field (left), the electric PE of an electrically-charged particle transforms to other forms of energy when it changes location in an electric field (right). When released, how does the KE acquired by a charged particle compare with the decrease in PE?

_____

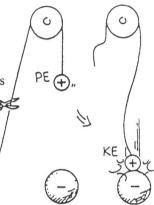

2. *Complete the statements.*

 A force compresses the spring. The work done in compression is the product of the average force and the distance moved. **W = Fd**. This work increases the PE of the spring.

Similarly, a force pushes the charge (call it a test charge) closer to the charged sphere. The work done in moving the test charge

is the product of the average _____ and the _____ moved.

W = _____. This work _____ the PE of the test charge.

If the test charge is released, it will be repelled and fly past the starting point. Its gain in KE at this

point is _____ to its decrease in PE.

At any point, a greater quantity of test charge means a greater amount of PE, but not a greater amount of PE *per quantity* of charge. The quantities PE (measured in joules) and PE/charge (measured in volts) are different concepts.

By definition: **Voltage = PE/charge.** 1 volt = 1 joule/1coulomb.

3. *Complete the statements.*

4. When a charge of 1 C in an electric field has an electric PE of 1 J, it has a voltage of 1 V. When a charge of 2 C has an electric PE of 2 J, its voltage is = _____V.

5. If a conductor connected to the terminal of a battery has a potential of 12 volts, then each coulomb of charge on the conductor has a PE of _____ J.

6. If a charge of 1 C has a PE of 5000 J, its voltage is

   _____V.

7. If a charge of 0.001 C has a PE of 5 J, its voltage is

   _____V.

8. If a charge of 0.0001 C has a PE of 0.5 J, its voltage is

   _____V.

9. If a rubber balloon is charged to 5000 V, and the quantity of charge on the balloon is 1 millionth coulomb, (0.000001 C) then the PE of this charge is only _____ J.

10. Some people get mixed up between force and pressure. Recall that pressure is force *per area*. Similarly, some people get mixed up between electric PE and voltage. According to this chapter, voltage is electric PE *per* _____.

> Experiment, not philosophical discussion, decides what is correct in science.

CONCEPTUAL PHYSICAL SCIENCE **EXPLORATIONS**

Chapter 11 Electricity
Ohm's Law

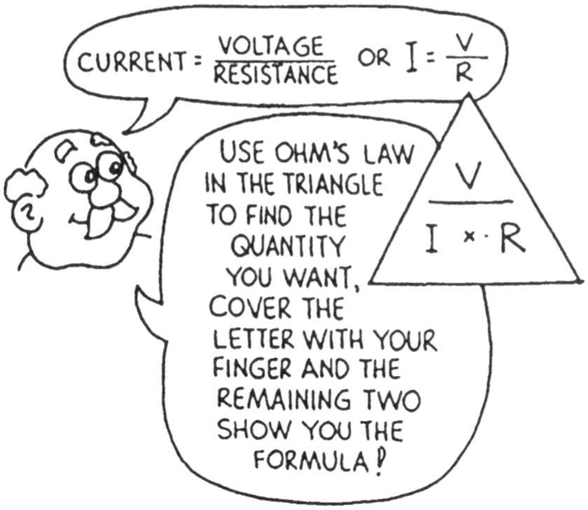

1. How much current flows in a 1000-ohm resistor when 1.5 volts are impressed across it?

   _____

2. If the filament resistance in an automobile headlamp is 3 ohms, how many amps does it draw when connected to a 12-volt battery?

   _____

3. The resistance of the side lights on an automobile are 10 ohms. How much current flows in them when connected to 12 volts?

   _____

4. What is the current in the 30-ohm heating coil of a coffee maker that operates on a 120-volt circuit?

   _____

5. During a lie detector test, a voltage of 6 V is impressed across two fingers. When a certain question is asked, the resistance between the fingers drops from 400,000 ohms to 200,000 ohms. What is the current (a) initially through the fingers, and (b) when the resistance between them drops?

   (a) _____ (b) _____

6. How much resistance allows an impressed voltage of 6 V to produce a current of 0.006 A?

   _____

7. What is the resistance of a clothes iron that draws a current of 12 A at 120 V?

   _____

8. What is the voltage across a 100-ohm circuit element that draws a current of 1 A?

   _____

9. What voltage will produce 3 A through a 15-ohm resistor?

   _____

10. The current in an incandescent lamp is 0.5 A when connected to a 120-V circuit, and 0.2 A when connected to a 10-V source. Does the resistance of the lamp change in these cases? Explain your answer and defend it with numerical values.

   _____

   _____

Chapter 11 Electricity
Electric Power

The rate that energy is converted from one form to another is *power*.

$$\text{power} = \frac{\text{energy converted}}{\text{time}} = \frac{\text{voltage} \times \text{charge}}{\text{time}} = \text{voltage} \times \frac{\text{charge}}{\text{time}} = \text{voltage} \times \text{current}$$

The unit of power is the *watt* (or *kilowatt*). So in units form,

Electric power (*watts*) = current (*amperes*) x voltage (*volts*),

where 1 *watt = 1 ampere x 1 volt* .

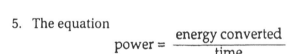

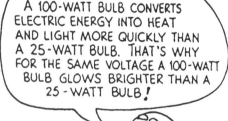

1. What is the power when a voltage of 120 V drives a 2-A current through a device?

   _____

2. What is the current when a 60-W lamp is connected to 120 V?

   _____

3. How much current does a 100-W lamp draw when connected to 120 V?

   _____

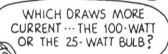

4. If part of an electric circuit dissipates energy at 6 W when it draws a current of 3 A, what voltage is impressed across it?

   _____

5. The equation
   $$\text{power} = \frac{\text{energy converted}}{\text{time}}$$
   rearranged gives

   energy converted = _____

6. Explain the difference between a kilowatt and a kilowatt-hour.

   _____

7. For safety reasons, some people leave their front porch light on all the time. If they use a 60-W bulb at 120 V, and the local power utility sells energy at 15 cents per kilowatt-hour, how much does it cost to leave the bulb on for a month? Show your work on the other side of this page.

CONCEPTUAL PHYSICAL SCIENCE **EXPLORATIONS**
Chapter 11 Electricity
Series Circuits

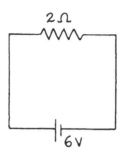

1. In the circuit shown at the right, a voltage of 6 V pushes charge through a single resistor of 2 Ω. According to Ohm's law, the current in the resistor (and therefore in the whole circuit) is

   _____ A.

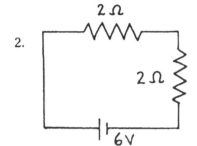

2. If a second identical lamp is added, as on the left, the 6-V battery must push charge through a total resistance of _____ Ω. The current in the circuit is then _____ A.

3. The equivalent resistance of three 4-Ω resistors in series is _____ Ω.

4. Does current flow *through* a resistor, or *across* a resistor? _____

   Is voltage established *through* a resistor, or *across* a resistor? _____

5. Does current in the lamps occur simultaneously, or does charge flow first through one lamp, then the other, and finally the last in turn?

   _____

6. Circuits a and b below are identical with all bulbs rated at equal wattage (therefore equal resistance). The only difference between the circuits is that Bulb 5 has a short circuit, as shown.

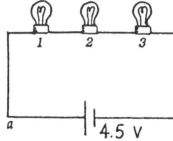 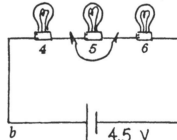

   a. In which circuit is the current greater? _____

   b. In which circuit are all three bulbs equally bright? _____

   c. What bulbs are the brightest? _____

   d. What bulb is the dimmest? _____

   e. What bulbs have the largest voltage drops across them? _____

   f. Which circuit dissipates more power? _____

   g. What circuit produces more light? _____

Chapter 11 Electricity

Parallel Circuits

1. In the circuit shown below, there is a voltage drop of
   6 V across *each* 2-Ω resistor.

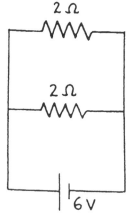

a. By Ohm's law, the current in each
   resistor is _____ A.

b. The current through the battery is
   the sum of the currents in the
   resistors, _____ A.

c. Fill in the current in the eight blank
   spaces in the view of the *same circuit*
   shown again at the right.

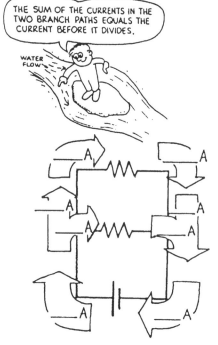

THE SUM OF THE CURRENTS IN THE
TWO BRANCH PATHS EQUALS THE
CURRENT BEFORE IT DIVIDES.

WATER FLOW

2. Cross out the circuit below that is not equivalent to
   the circuit above.

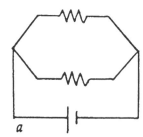

a

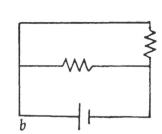

b

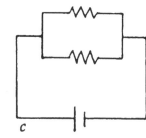

c

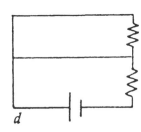

d

3. Consider the parallel circuit at the right.
   a. The voltage drop across each resistor is
      _____ V.

   b. The current in each branch is:

   2-Ω resistor _____ A

   2-Ω resistor _____ A

   1-Ω resistor _____ A

   b. The current through the battery
      equals the sum of the currents which
      equals _____ A.

   c. The equivalent resistance of the circuit
      equals _____ Ω.

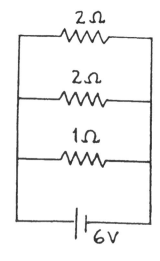

THE EQUIVALENT
RESISTANCE OF A
PAIR OF RESISTORS
IN PARALLEL IS
THEIR PRODUCT
DIVIDED BY
THEIR SUM!

Name Period Date

CONCEPTUAL PHYSICAL SCIENCE **EXPLORATIONS**

Chapter 12 Magnetism

Magnetism

Fill in each blank with the appropriate word.

1. Attraction or repulsion of charges depends on their *signs*, positives or negatives. Attraction or repulsion of magnets depends on their magnetic ____________: ____________ or ____________.

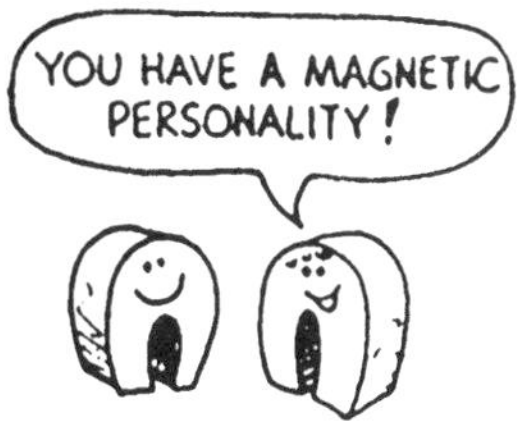

YOU HAVE A MAGNETIC PERSONALITY!

2. Opposite poles attract; like poles __________.

3. A magnetic field is produced by the __________ of electric charge.

4. Clusters of magnetically aligned atoms are magnetic ____________________.

5. A magnetic __________ surrounds a current-carrying wire.

6. When a current-carrying wire is made to form a coil around a piece of iron, the result is an ____.

7. A charged particle moving in a magnetic field experiences a deflecting ________________ that is maximum when the charge moves ____________________ to the field.

S
N

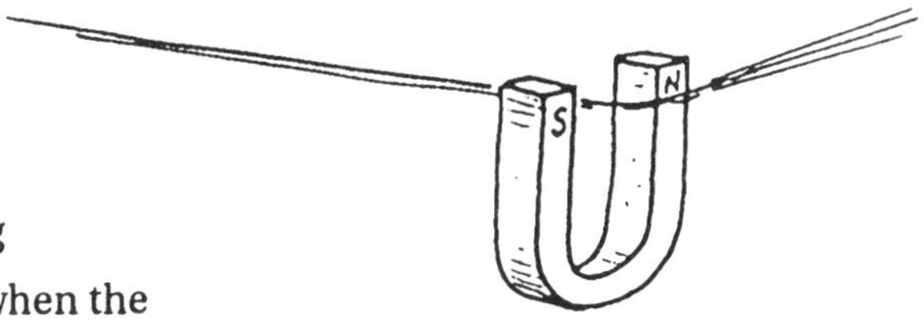

8. A current-carrying wire experiences a deflecting ______________________ that is maximum when the wire and magnetic field are ______________________ to one another.

9. A simple instrument designed to detect electric current is the ______________________; when calibrated to measure current, it is an ______________________; when calibrated to measure voltage, it is a ______________________.

10. The largest size magnet in the world is the ______________ itself.

THEN TO REALLY MAKE THINGS "SIMPLE," THERE'S THE RIGHT-HAND RULE!
F
B
I

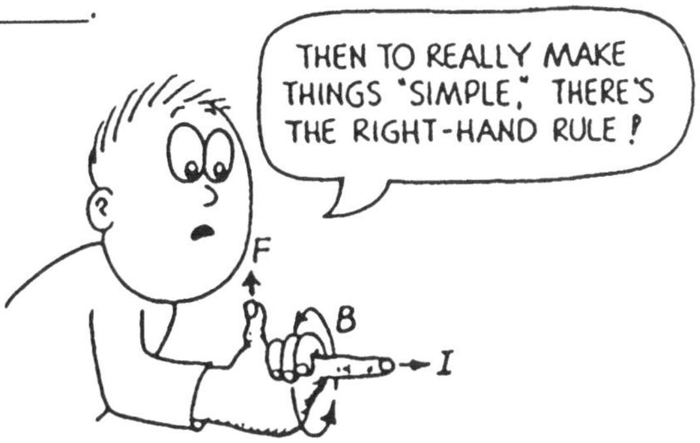

11. The illlustration below is similar to Figure 12.2 in your textbook. Iron filings trace out patterns of magnetic field lines about a bar magnet. In the field are some magnetic compasses. The compass needle in only one compass is shown. Draw in the needles with proper orientation in the other compasses.

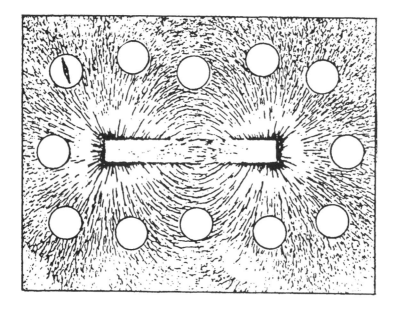

12. The illustration below is similar to Figure 12.12 (center) in your textbook. Iron filings trace out the magnetic field pattern about the loop of current-carrying wire. Draw in the compass needle orientations for all the compasses.

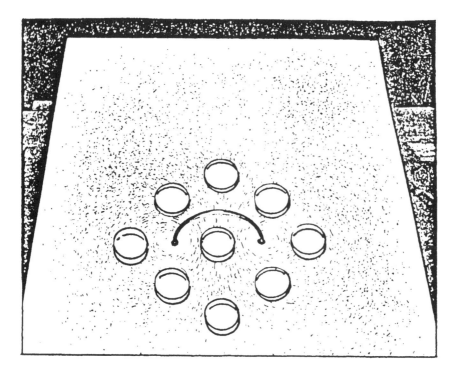

CONCEPTUAL PHYSICAL SCIENCE **EXPLORATIONS**

Chapter 12 Magnetism

Faraday's Law

1. Hans Christian Oersted discovered that magnetism and electricity are

    (related)   (independent of each other).

    Magnetism is produced by

    (batteries)   (the motion of electric charges).

    Faraday and Henry discovered that electric current can be produced by

    (batteries)   (motion of a magnet).

    More specifically, voltage is induced in a loop of wire if there is a change in the

    (batteries)   (magnetic field in the loop).

    This phenomenon is called

    (electromagnetism)   (electromagnetic induction).

2. When a magnet is plunged in and out of a coil of wire, voltage is induced in the coil. If the rate of the in-and-out motion of the magnet is doubled, the induced voltage

    (doubles)   (halves)   (remains the same).

    If instead the number of loops in the coil is doubled, the induced voltage

    (doubles)   (halves)   (remains the same).

3. A rapidly changing magnetic field in any region of space induces a rapidly changing

    (electric field)   (**magnetic field**)   (gravitational field)

    which in turn induces a rapidly changing

    (magnetic field)   (electric field)   (baseball field).

    This generation and regeneration of electric and magnetic fields make up

    (electromagnetic waves)   (sound waves)   (both of these).

CONCEPTUAL PHYSICAL SCIENCE **EXPLORATIONS**

Chapter 13 Waves and Sound
Vibrations and Waves

1. A sine curve that represents a transverse wave is drawn below. With a ruler, measure the wavelength and amplitude of the wave.

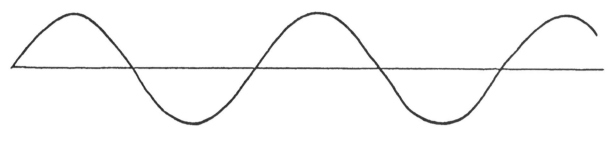

a. Wavelength = _____   b. Amplitude = _____

2. A kid on a playground swing makes a complete to-and-fro swing each 2 seconds. The frequency of swing is

    (0.5 hertz)   (1 hertz)   (2 hertz)

    and the period is

    (0.5 second)   (1 second)   (2 seconds)

3. *Complete the statements.*

THE PERIOD OF A 440-HERTZ SOUND WAVE IS _____ SECOND.

A MARINE WEATHER STATION REPORTS WAVES ALONG THE SHORE THAT ARE 8 SECONDS APART. THE FREQUENCY OF THE WAVES IS THEREFORE _____ HERTZ.

4. The annoying sound from a mosquito is produced when it beats its wings at the average rate of 600 wingbeats per second.

    a. What is the frequency of the sound waves?

    _____

    b. What is the wavelength? (Assume the speed of sound is 340 m/s.)

    _____

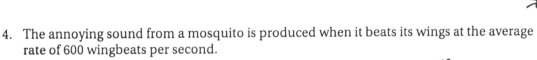

Vibrations and Waves continued:

5. A rapid-fire pellet gun fires 10 rounds per second.
   The speed of the pellets is 300 m/s.

   a. What is the distance in the air between the flying pellets?_____

   b. What happens to the distance between the pellets if the rate of fire is increased?

   _____

6. Consider a wave generator that produces 10 pulses per second. The speed of the waves is 300 cm/s.

   a. What is the wavelength of the waves? _____

   b. What happens to the wavelength if the frequency of pulses is increased?

   _____

7. The bird at the right watches the waves. If the portion of a wave between 2 crests passes the pole each second, what is the speed of the wave?

   _____

   What is its period?

   _____

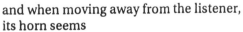

8. If the distance between crests in the above question were 1.5 meters apart, and 2 crests pass the pole each second, what would be the speed of the wave?

   _____

   What would be its period?

   _____

9. When an automobile moves toward a listener, the sound of its horn seems relatively

   (low pitched)   (normal)

   (high pitched)

   and when moving away from the listener, its horn seems

   (low pitched)   (normal)

   (high pitched)

10. The changed pitch of the Doppler effect is due to changes in

    (wave speed) (wave frequency)

CONCEPTUAL PHYSICAL SCIENCE **EXPLORATIONS**

Chapter 13 Waves and Sound
Sound

1. Two major classes of waves are *longitudinal* and *transverse*. Sound waves are

   (longitudinal)   (transverse)

   > Sound is the only thing we hear!

2. The frequency of a sound signal refers to how frequently the vibrations occur. A high-frequency sound is heard at a high

   (pitch)   (wavelength)   (speed)

3. The sketch below shows a snapshot of the compressions and rarefactions of the air in a tube as the sound moves toward the right. The dots represent molecules. With a ruler the wavelength of the sound wave is measured to be _____ cm.

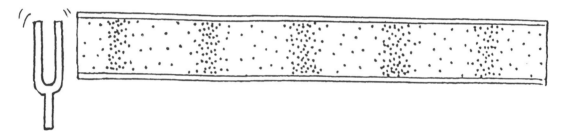

4. Compared to the wavelengths of high-pitched sounds, the wavelengths of low-pitched sounds are

   (long)   (short)

5. Suppose you set your watch by the sound of the noon whistle from a factory 3 km away.

   a. Compared to the correct time, your watch will be
      (behind)   (ahead)

   b. It will differ from the correct time by
      (3 seconds)   (6 seconds)   (9 seconds)

LET'S SEE, FROM $v = \dfrac{d}{t}$

$t = \dfrac{d}{v} = \dfrac{3000 \, m}{340 \, m/s} = \ldots$

PHYSICS PHYSICS

Sound continued:

6. Sound waves travel fastest in

    (solids) (liquids) (gases)

    (...same speed in each)

7. If the child's natural frequency of swinging is once each 4 seconds, for maximum amplitude the man should push at a rate of once each

    (2 seconds) (4 seconds) (8 seconds)

8. If the man in Question 7 pushes in the same direction twice as often, his pushes

    (will) (will not)

be effective because

    (the swing will be pushed twice as often in the right direction)

    (every other push will oppose the motion of the swing)

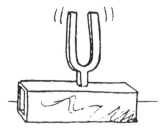

9. The frequency of the tuning fork is 440 hertz. It will NOT be forced into vibration by a sound of

    (220 hertz) (440 hertz) (880 hertz)

10. Beats are the result of the alternate cancellation and reinforcement of two sound waves of

    (the same frequency) (slightly different frequencies)

11. Beat frequency equals the difference between frequencies. So if two notes with frequencies of 66 and 70 Hz aresounded together, the resulting beat frequency is

    (4 hertz) (68 hertz) (136 hertz)

12. The accepted value for the speed of sound in air is 332 m/s at 0°C. The speed of sound in air increases 0.6 m/s for each Celsius degree above zero. Compute the speed of sound at the temperature of the room you are now in.

CONCEPTUAL PHYSICAL SCIENCE **EXPLORATIONS**

Chapter 13 Waves and Sound

Shock Waves

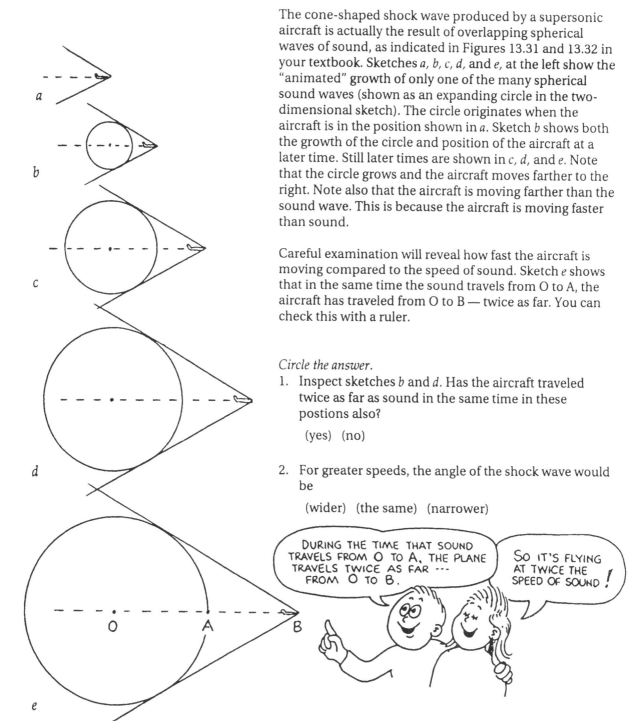

The cone-shaped shock wave produced by a supersonic aircraft is actually the result of overlapping spherical waves of sound, as indicated in Figures 13.31 and 13.32 in your textbook. Sketches *a, b, c, d,* and *e,* at the left show the "animated" growth of only one of the many spherical sound waves (shown as an expanding circle in the two-dimensional sketch). The circle originates when the aircraft is in the position shown in *a.* Sketch *b* shows both the growth of the circle and position of the aircraft at a later time. Still later times are shown in *c, d,* and *e.* Note that the circle grows and the aircraft moves farther to the right. Note also that the aircraft is moving farther than the sound wave. This is because the aircraft is moving faster than sound.

Careful examination will reveal how fast the aircraft is moving compared to the speed of sound. Sketch *e* shows that in the same time the sound travels from O to A, the aircraft has traveled from O to B — twice as far. You can check this with a ruler.

Circle the answer.

1. Inspect sketches *b* and *d*. Has the aircraft traveled twice as far as sound in the same time in these postions also?

   (yes) (no)

2. For greater speeds, the angle of the shock wave would be

   (wider) (the same) (narrower)

Shock Waves continued:

3. Use a ruler to estimate the speeds of the aircraft that produce the shock waves in the two sketches below.

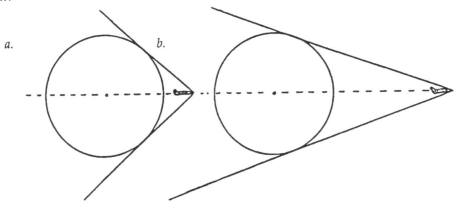

a.

b.

Aircraft *a* is traveling about _____ times the speed of sound.

Aircraft *b* is traveling about _____ times the speed of sound.

4. Draw your own circle (anywhere) and estimate the speed of the aircraft to produce the shock wave shown below. It will be helpful to use a coin to draw your circle. Then move it and count the number of diameters that make up your shock wave.

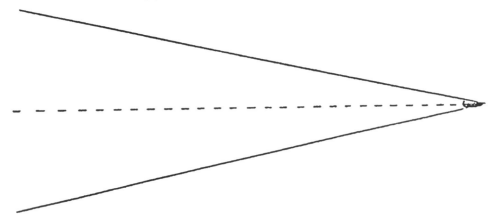

The speed is about _____ times the speed of sound.

5. In the space below, draw the shock wave made by a supersonic missile that travels at four times the speed of sound.

CONCEPTUAL PHYSICAL SCIENCE **EXPLORATIONS**

Chapter 14 Light and Color
Light

1. Study Figure 14.3 in your textbook and answer the following:

   a. Which has the longer *wavelengths,* radio waves or light waves?

   _____

   b. Which has the longer *wavelengths,* light waves or gamma rays?

   _____

   c. Which has the higher *frequencies,* ultraviolet or infrared waves?

   _____

   d. Which has the higher *frequencies,* ultraviolet waves or gamma rays?

   _____

2. Carefully study Section 14.2 in your textbook and answer the following:

   a. Exactly what do vibrating electrons emit?

   _____

   b. When ultraviolet light shines on glass, what does it do to electrons in the glass structure?

   _____

   c. When energetic electrons in the glass structure vibrate against neighboring atoms, what happens to the energy of vibration?

   _____

   d. What happens to the energy of a vibrating electron that does not collide with neighboring atoms?

   _____

   e. Which range of light frequencies, visible or ultraviolet, are absorbed in glass?

   _____

   f. Which range of light frequencies, visible or ultraviolet, are transmitted through glass?

   _____

   g. How is the speed of light in glass affected by the succession of time delays that accompany the absorption and re-emission of light from atom to atom in the glass?

   _____

   h. How does the speed of light compare in water, glass, and diamond?

   _____

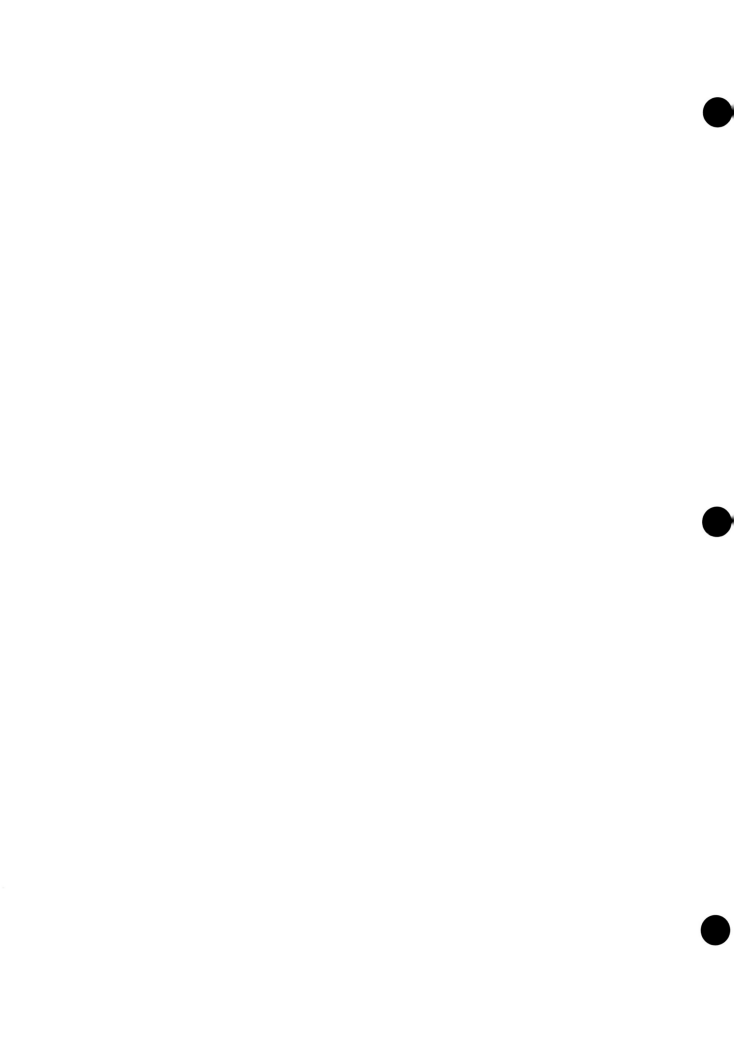

CONCEPTUAL PHYSICAL SCIENCE **EXPLORATIONS**

Chapter 14 Light and Color
Color

The sketch to the right shows the shadow of a teacher in front of a white screen in a darkened room. The light source is red, so the screen looks red and the shadow looks black. Color the sketch with colored markers, or label the colors with pen or pencil.

A green lamp is added and makes a second shadow. The formerly black shadow cast by the red light is no longer black, but is illuminated with green light. So it is green. Color or mark it green. The shadow cast by the green lamp is not black, because it is illuminated with the red light. Color or mark its color. The background receives a mixture of red and green light. Figure out what color the background will appear, then color or label it.

A blue lamp is added and three shadows appear. Color or label the appropriate colors of the shadows and the background.

The lamps are placed closer together so the shadows overlap. Indicate the colors of all screen areas.

Color continued:

If you have colored markers, have a go at these.

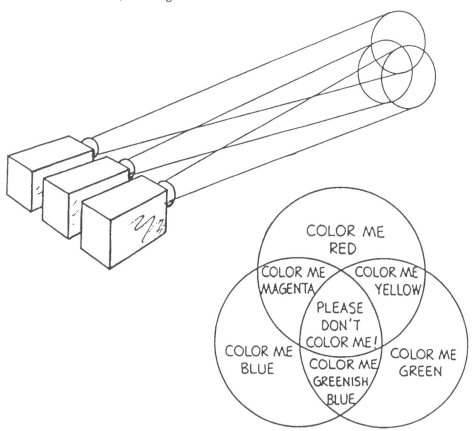

CONCEPTUAL PHYSICAL SCIENCE **EXPLORATIONS**

Chapter 15 Reflection and Refraction
Reflection

1. Light from a flashlight shines on a mirror and illuminates one of the cards. Draw the reflected beam to indicate the illuminated card.

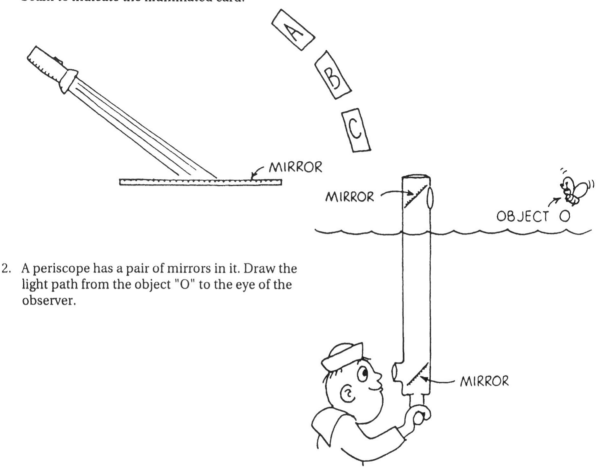

2. A periscope has a pair of mirrors in it. Draw the light path from the object "O" to the eye of the observer.

3. The ray diagram below shows the extension of one of the reflected rays from the plane mirror. Complete the diagram by (1) carefully drawing the three other reflected rays, and (2) extending them behind the mirror to locate the image of the flame. (Assume the candle and image are viewed by an observer on the left.)

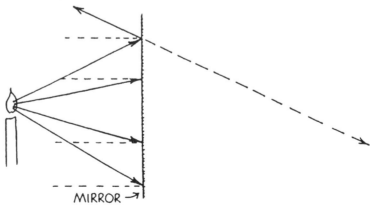

Reflection continued:

4. The ray diagram below shows the reflection of one of the rays that strikes the parabolic mirror. Notice that the law of reflection is obeyed, and the angle of incidence (from the normal, the dashed line) equals the angle of reflection (from the normal). Complete the diagram by drawing the reflected rays of the other three rays that are shown. (Do you see why parabolic mirrors are used in automobile headlights?)

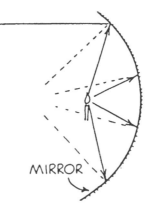

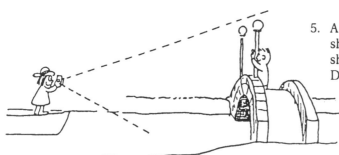

5. A girl takes a photograph of the bridge as shown. Which of the two sketches correctly shows the reflected view of the bridge? Defend your answer.

_____

_____

_____

_____

CONCEPTUAL PHYSICAL SCIENCE **EXPLORATIONS**

Chapter 15 Reflection and Refraction
Refraction

1. A pair of toy cart wheels are rolled obliquely from a smooth surface onto two plots of grass — a rectangular plot as shown at the left, and a triangular plot as shown at the right. The ground is on a slight incline so that after slowing down in the grass, the wheels speed up again when emerging on the smooth surface. Finish each sketch and show some positions of the wheels inside the plots and on the other side. Clearly indicate their paths and directions of travel.

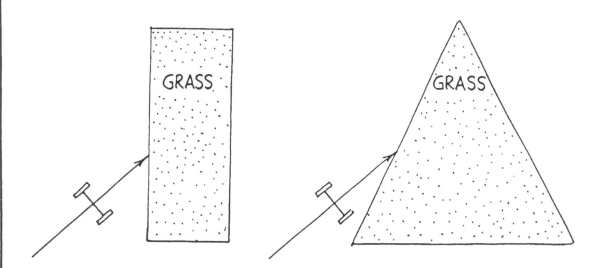

2. Red, green, and blue rays of light are incident upon a glass prism as shown. The average speed of red light in the glass is less than in air, so the red ray is refracted. When it emerges into the air it regains its original speed and travels in the direction shown. Green light takes longer to get through the glass. Because of its slower speed it is refracted as shown. Blue light travels even slower in glass. Complete the diagram by estimating the path of the blue ray.

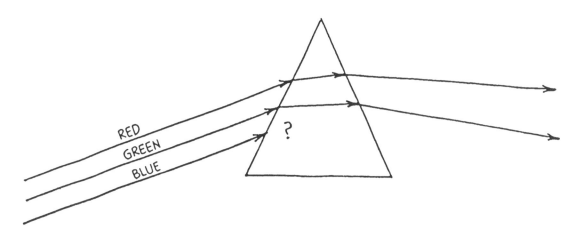

Refraction continued:

3. The sketch to the right shows a light ray moving from air into water, at 45° to the normal. Which of the three rays indicated with capital letters is most likely the light ray that continues inside the water?

_____

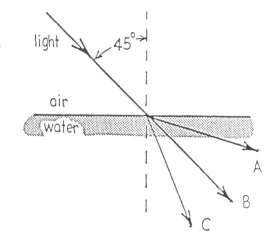

4. The sketch on the left shows a light ray moving from glass into air, at 30° to the normal. Which of the three is most likely the light ray that continues in the air?

_____

5. To the right, a light ray is shown moving from air into a glass block, at 40° to the normal. Which of the three rays is most likely the light ray that travels in the air after emerging from the opposite side of the block?

_____

Sketch the path the light would take inside the glass.

6. To the left, a light ray is shown moving from water into a rectangular block of air (inside a thin-walled plastic box), at 40° to the normal. Which of the three rays is most likely the light ray that continues into the water on the opposite side of the block?

_____

Sketch the path the light would take inside the air.

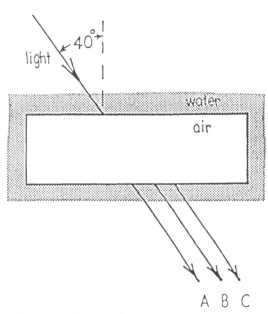

thanx to Clarence Bakken

CONCEPTUAL PHYSICAL SCIENCE **EXPLORATIONS**

Chapter 15 Reflection and Refraction
Measuring the Diameter of the Sun with a Ruler

Look carefully at the round spots of light on the shady ground beneath trees. These are *sunballs*, which are images of the sun. They are cast by openings between leaves in the trees that act as pinholes. (Did you make a pinhole "camera" back in middle school?) Large sunballs, several centimeters in diameter or so, are cast by openings that are relatively high above the ground, while small ones are produced by

closer "pinholes." The interesting point is that the ratio of the diameter of the sunball to its distance from the pinhole is the same as the ratio of the sun's diameter to its distance from the pinhole. We know the sun is approximately 150,000,000 km from the pinhole, so careful measurements of the ratio of diameter/distance for a sunball leads you to the diameter of the sun. That's what this page is about. Instead of measuring sunballs under the shade of trees on a sunny day, make your own easier-to-measure sunball.

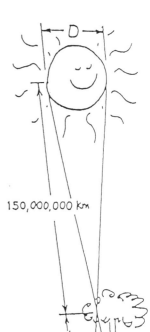

150,000,000 km

1. Poke a small hole in a piece of card. Perhaps an index card will do, and poke the hole with a sharp pencil or pen. Hold the card in the sunlight and note the circular image that is cast. This is an image of the sun. Note that its size doesn't depend on the size of the hole in the card, but only on its distance. The image is a circle when cast on a surface perpendicular to the rays — otherwise it's "stretched out" as an ellipse.

2. Try holes of various shapes; say a square hole, or a triangular hole. What is the shape of the image when its distance from the card is large compared with the size of the hole? Does the shape of the pinhole make a difference?

_____

_____

3. Measure the diameter of a small coin. Then place the coin on a viewing area that is perpendicular to the sun's rays. Position the card so the image of the sunball exactly covers the coin. Carefully measure the distance between the coin and the small hole in the card. Complete the following:

$$\frac{\text{Diameter of sunball}}{\text{Distance to pinhole}} = \underline{\hspace{2cm}}$$

> WHAT SHAPE DO SUNBALLS HAVE DURING A PARTIAL ECLIPSE OF THE SUN ?

With this ratio, estimate the diameter of the sun. Show your work on a separate piece of paper.

4. If you did this on a day when the sun is partially eclipsed, what shape of image would you expect to see?

_____

Chapter 15 Reflection and Refraction
Lenses

Rays of light bend as shown when passing through the glass blocks.

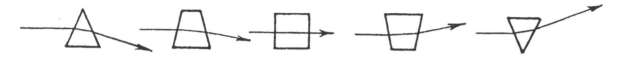

1. Show how light rays bend when they pass through the arrangement of glass blocks shown below.

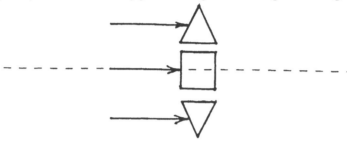

2. Show how light rays bend when they pass through the lens shown below. Is the lens a converging or a diverging lens? What is your evidence?

_____

_____

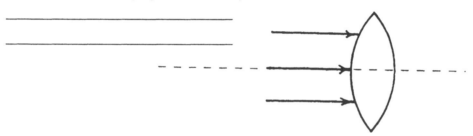

3. Show how light rays bend when they pass through the arrangement of glass blocks shown below.

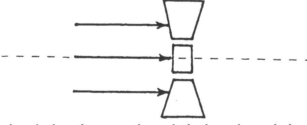

4. Show how light rays bend when they pass through the lens shown below. Is the lens a converging or a diverging lens? What is your evidence?

_____

_____

Being able to see is wonderful—not to be taken lightly!

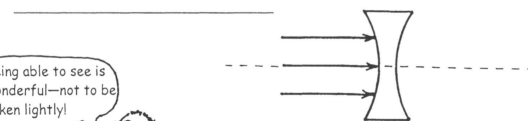

CONCEPTUAL PHYSICAL SCIENCE **EXPLORATIONS**
Chapter 16 Properties of Light
Interference

1. Sketch an identical wave upon the one below, that is in phase. Then sketch the resulting wave.

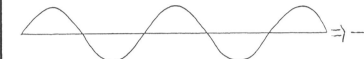

2. Here is the wave again. This time sketch an identical wave upon it that is out of phase. Sketch the result.

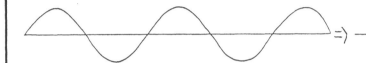

3. Figure 16.9 from your text is repeated below. Carefully count the number of wavelengths (same as the number of wave crests) along the following paths between the slits and the screen.

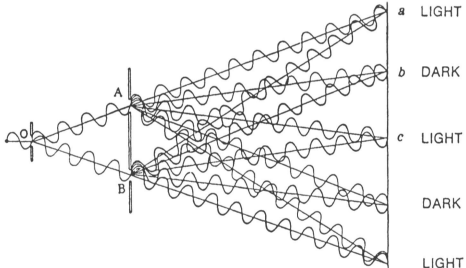

a. Number of wavelengths between slit A and point a = _____

b. Number of wavelengths between slit B and point *a* = _____

c. Number of wavelengths between slit A and point *b* = _____

d. Number of wavelengths between slit B and point *b* = _____

e. Number of wavelengths between slit A and point *c* = _____

f. Number of wave crests between slit B and point *c* = _____

When the number of wavelengths along each path is the same or differs by one or more whole wavelengths, interference is (constructive) (destructive)

and when the number of wavelengths differ by a half wavelength (or odd multiples of a half wavelength), interference is (constructive) (destructive).

CONCEPTUAL PHYSICAL SCIENCE EXPLORATIONS

Chapter 16 Properties of Light
Polarization

The amplitude of a light wave has magnitude and direction, and can be represented by a vector. Polarized light vibrates in a single direction and is represented by a single vector. To the left the single vector represents vertically polarized light. The vibrations of nonpolarized light are equal in all directions. There are as many vertical components as horizontal components. The pair of perpendicular vectors to the right represents nonpolarized light.

1. In the sketch below nonpolarized light from a flashlight strikes a pair of Polaroid filters.

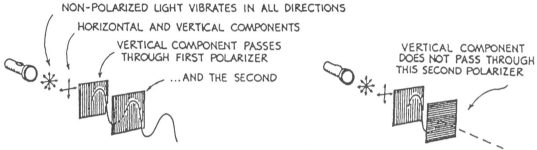

NON-POLARIZED LIGHT VIBRATES IN ALL DIRECTIONS
HORIZONTAL AND VERTICAL COMPONENTS
VERTICAL COMPONENT PASSES THROUGH FIRST POLARIZER
...AND THE SECOND

VERTICAL COMPONENT DOES NOT PASS THROUGH THIS SECOND POLARIZER

 a. Light is transmitted by a pair of Polaroids when their axes are

   (aligned)   (crossed at right angles)

   and light is blocked when their axes are

   (aligned)   (crossed at right angles)

 b. Transmitted light is polarized in a direction

   (the same as)   (different than)   the polarization axis of the filter.

2. Consider the transmission of light through a pair of Polaroids with polarization axes at 45° to each other. Although in practice the Polaroids are one atop the other, we show them spread out side by side below. From left to right: (*a*) Nonpolarized light is represented by its horizontal and vertical components. (*b*) These components strike filter A. (*c*) The vertical component is transmitted, and (*d*) falls upon filter B. This vertical component is not aligned with the polarization axis of filter B, but it has a component that is — component *t*; (*e*) which is transmitted.

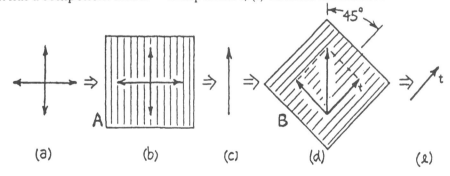

(a)   (b)   (c)   (d)   (e)

 a. The amount of light that gets through Filter B, compared to the amount that gets through Filter A is

   (more) (less) (the same)

 b. The component perpendicular to *t* that falls on Filter B is

   (also transmitted) (absorbed)

Chapter 16 Properties of Light

Wave-Particle Duality

1. To say that light is quantized means that light is made up of

    (elemental units)

    (waves)

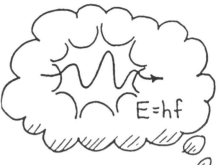

2. Compared to photons of low-frequency light, photons of higher-frequency light have more

    (energy)

    (speed)

    (quanta)

3. The photoelectric effect supports the

    (wave model of light)

    (particle model of light)

4. The photoelectric effect is evident when light shone on certain photosensitive materials ejects

    (photons)

    (electrons)

5. 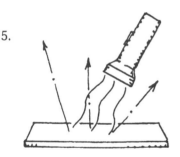 The photoelectric effect is more effective with violet light than with red light because the photons of violet light

    (resonate with the atoms in the material)

    (deliver more energy to the material)

    (are more numerous)

6. According to the wave-particle model of light, light travels as a

    (wave and hits like a particle)   (particle and hits like a wave)

7. According to the wave-particle model of matter, a beam of light and a beam of electrons

    (are fundamentally different)   (are similar)

8. Whereas in the everyday macroworld the study of motion is called *mechanics*, in the microworld the study of quanta is called

    (Newtonian mechanics)   (quantum mechanics)

CONCEPTUAL PHYSICAL SCIENCE **EXPLORATIONS**

Chapter 17 Atoms and the Periodic Table
Subatomic Particles

Three fundamental particles of the atom are the _____, _____, and _____ .

At the center of each atom lies the atomic _____, which consists of _____ and

_____ . The **atomic number** refers to the number of _____ in the nucleus. All

atoms of the same element have the same number of _____, hence, the same atomic

number.

Isotopes are atoms that have the same number of _____ but a different number of

_____ . An isotope is identified by its **atomic mass number**, which is the total number of

_____ and _____ in the nucleus. A carbon isotope that has 6 _____ and

6 _____ is identified as carbon-12, where 12 is the atomic mass number. A carbon isotope

having 6 _____and 8 _____, on the other hand, is carbon-14.

1. *Complete the following table:*

Isotope	Number of...		
	Electrons	Protons	Neutrons
Hydrogen-1	1		
Chlorine-36		17	
Nitrogen-14			7
Potassium-40	19		
Arsenic-75		33	
Gold-197			118

2. Which results in a more valuable product — *adding* or *subtracting* protons from gold nuclei?

_____

3. Which has more mass, a helium atom or a neon atom?

_____

4. Which has a greater number of atoms, a gram of
   helium or a gram of neon?

_____

CONCEPTUAL PHYSICAL SCIENCE **EXPLORATIONS**

Chapter 17 Atoms and the Periodic Table
Melting Points of the Elements

There is a remarkable degree of organization in the periodic table. As discussed in your textbook, elements within the same atomic group (vertical column) share similar properties. Also, the chemical reactivity of an element can be deduced from its position in the periodic table. Two additional examples of the periodic table's organization are the melting points and densities of the elements.

The periodic table below shows the melting points of nearly all the elements. Note the melting points are not randomly oriented, but, with only a few exceptions, either gradually increase or decrease as you move in any particular direction. This can be clearly illustrated by color coding each element according to its melting point.

Use colored pencils to color in each element according to its melting point. Use the suggested color legend. Color lightly so that symbols and numbers are still visible.

Color	Temperature Range, °C	Color	Temperature Range, °C
Violet	-273 — -50	Yellow	1400 — 1900
Blue	-50 — 300	Orange	1900 — 2900
Cyan	300 — 700	Red	2900 — 3500
Green	700 — 1400		

| 1 | 2 | 3 | 4 | 5 | 6 | 7 | 8 | 9 | 10 | 11 | 12 | 13 | 14 | 15 | 16 | 17 | 18 |

Melting Points of the Elements (°C)

H -259																	He -272
Li 180	Be 1278											B 2079	C 3550	N -210	O -218	F -219	Ne -248
Na 97	Mg 648											Al 660	Si 1410	P 44	S 113	Cl -100	Ar -189
K 63	Ca 839	Sc 1541	Ti 1660	V 1890	Cr 1857	Mn 1244	Fe 1535	Co 1495	Ni 1453	Cu 1083	Zn 419	Ga 30	Ge 937	As 817	Se 217	Br -7	Kr -156
Rb 39	Sr 769	Y 1522	Zr 1852	Nb 2468	Mo 2617	Tc 2172	Ru 2310	Rh 1966	Pd 1554	Ag 961	Cd 320	In 156	Sn 231	Sb 630	Te 449	I 113	Xe -111
Cs 28	Ba 725	La 921	Hf 2227	Ta 2996	W 3410	Re 3180	Os 3045	Ir 2410	Pt 1772	Au 1064	Hg -38	Tl 303	Pb 327	Bi 271	Po 254	At 302	Rn -71
Fr 27	Ra 700	Ac 1050	--	--	--	--	--	--									

Lanthanides:	Ce 799	Pr 931	Nd 1021	Pm 1168	Sm 1077	Eu 822	Gd 1313	Tb 1356	Dy 1412	Ho 1474	Er 1159	Tm 1545	Yb 819	Lu 1663

Actinides:	Th 1750	Pa 1600	U 1132	Np 640	Pu 641	Am 994	Cm 1340	Bk --	Cf --	Es --	Fm --	Md --	No --	Lr --

1. Which elements have the highest melting points?

_____

2. Which elements have the lowest melting points?

_____

3. Which atomic groups tend to go from higher to lower melting points reading from top to bottom? (Identify each group by its group number).

_____

4. Which atomic groups tend to go from lower to higher melting points reading from top to bottom?

_____

Densities of the Elements

The periodic table below shows the densities of nearly all the elements. As with the melting points, the densities of the elements either gradually increase or decrease as you move in any particular direction. Use colored pencils to color in each element according to its density. Shown below is a suggested color legend. Color lightly so that symbols and numbers are still visible. (Note: All gaseous elements are marked with an asterisk and should be the same color. Their densities, which are given in units of g/L, are much less than the densities non-gaseous elements, which are given in units of g/mL.)

Color	Density (g/mL)	Color	Density (g/mL)
Violet	gaseous elements	Yellow	16 — 12
Blue	5 — 0	Orange	20 — 16
Cyan	8 — 5	Red.	23 — 20
Green	12 — 8		

1	2	3	4	5	6	7	8	9	10	11	12	13	14	15	16	17	18
H * 0.09																	He * 0.18
Li 0.5	Be 1.8											B 2.3	C 2.0	N * 1.25	O * 1.43	F * 1.70	Ne * 0.90
Na 1.0	Mg 1.7											Al 2.7	Si 2.3	P 1.8	S 2.1	Cl * 3.21	Ar * 1.78
K 0.9	Ca 1.6	Sc 3.0	Ti 4.5	V 6.1	Cr 7.2	Mn 7.3	Fe 7.8	Co 8.9	Ni 8.9	Cu 9.0	Zn 7.1	Ga 6.1	Ge 5.3	As 5.7	Se 4.8	Br * 7.59	Kr * 3.73
Rb 1.5	Sr 2.5	Y 4.5	Zr 6.5	Nb 8.5	Mo 6.8	Tc 11.5	Ru 12.4	Rh 12.4	Pd 12.0	Ag 10.5	Cd 8.7	In 7.3	Sn 5.7	Sb 6.7	Te 6.2	I 4.9	Xe * 5.89
Cs 1.9	Ba 3.5	La 6.2	Hf 13.3	Ta 16.6	W 19.3	Re 21.0	Os 22.6	Ir 22.4	Pt 21.5	Au 18.9	Hg 13.5	Tl 11.9	Pb 11.4	Bi 9.7	Po 9.3	At --	Rn * 9.73
Fr --	Ra 5.0	Ac 10.1	Unq --	Unp --	Unh --	Uns --	Uno --	Une --									

Densities of the Elements (g/mL)

* density of gaseous phase in g/L

Lanthanides:	Ce	Pr	Nd	Pm	Sm	Eu	Gd	Tb	Dy	Ho	Er	Tm	Yb	Lu
	6.7	6.7	6.8	7.2	7.5	5.2	7.9	8.2	8.6	8.8	9.1	9.3	6.9	9.8

Actinides:	Th	Pa	U	Np	Pu	Am	Cm	Bk	Cf	Es	Fm	Md	No	Lr
	11.7	15.4	19.0	20.1	19.8	13.7	13.5	14	--	--	--	--	--	--

1. Which elements are the most dense?

_____

2. How variable are the densities of the lanthanides compared to the densities of the actinides?

_____

3. Which atomic groups tend to go from higher to lower densities reading from top to bottom?
   (Identify each group by its group number).

_____

4. Which atomic groups tend to go from lower to higher densities reading from top to bottom?

_____

CONCEPTUAL PHYSICAL SCIENCE **EXPLORATIONS**

Chapter 18 Atomic Models
Losing Valence Electrons

The shell model described in Section 18.4 can be used to explain a wide variety of properties of atoms. Using the shell model, for example, we can explain how atoms within the same group tend to lose (or gain) the same number of electrons. Let's consider the case of three group 1 elements: lithium, sodium, and potassium. Look to a periodic table and find the nuclear charge of each of these atoms:

   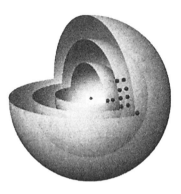   

Lithium, Li   Sodium, Na   Potassium, K

Nuclear
charge: _____   _____   _____

Number of
inner shell
electrons: _____   _____   _____

How strongly the valence electron is held to the nucleus depends on the strength of the nuclear charge—the stronger the charge, the stronger the valence electron is held. There's more to it, however, because inner-shell electrons weaken the attraction outer-shell electrons have for the nucleus. The valence shell in lithium, for example, doesn't experience the full effect of three protons. Instead, it experiences a diminished nuclear charge of about +1. We get this by subtracting the number of inner-shell electrons from the actual nuclear charge. What do the valence electrons for sodium and potassium experience?

Diminished   **about**
nuclear   **+1**
charge: _____   _____   _____

Question: Potassium has a nuclear charge many times greater than that of lithium. Why is it actually *easier* for a potassium atom to lose its valence electron than it is for a lithium atom to lose its valence electron?

Hint: Remember from Chapter 11
what happens to the electric force
as distance is increased!

_____

_____

_____

_____

CONCEPTUAL PHYSICAL SCIENCE **EXPLORATIONS**
Chapter 19 Radioactivity
The Atomic Nucleus and Radioactivity

1. *Complete the following statements.*

    a. A lone neutron spontaneously decays into a proton plus an

        _____ .

    b. Alpha and beta rays are made of streams of particles, whereas
        gamma rays are streams of _____ .

    c. An electrically charged atom is called an_____.

    d. Different _____of an element are chemically identical but differ in the number
        of neutrons in the nucleus.

    e. Transuranic elements are those beyond atomic number_____.

    f. If the amount of a certain radioactive sample decreases by half in four weeks, in four more
        weeks the amount remaining should be _____the original amount.

    g. Water from a natural hot spring is warmed by_____inside the Earth.

2. The gas in the little girl's balloon is made up of former alpha and beta particles produced by
    radioactive decay.

    a. If the mixture is electrically neutral, how many more beta
        particles than alpha particles are in the balloon?

        _____

        _____

    b. Why is your answer not "same"?

        _____

        _____

    c. Why are the alpha and beta particles no longer harmful to the child?

        _____

        _____

    d. What element does this mixture make?

        _____

Chapter 19 Radioactivity

Natural Transmutation

Fill in the decay-scheme diagram below, similar to that shown in **Figure** 19.11 in the textbook, but beginning with U-235 and ending up with an isotope of lead. Use the table at the left, and identify each element in the series with its chemical symbol.

Step	Particle emitted
1	Alpha
2	Beta
3	Alpha
4	Alpha
5	Beta
6	Alpha
7	Alpha
8	Alpha
9	Beta
10	Alpha
11	Beta
12	Stable

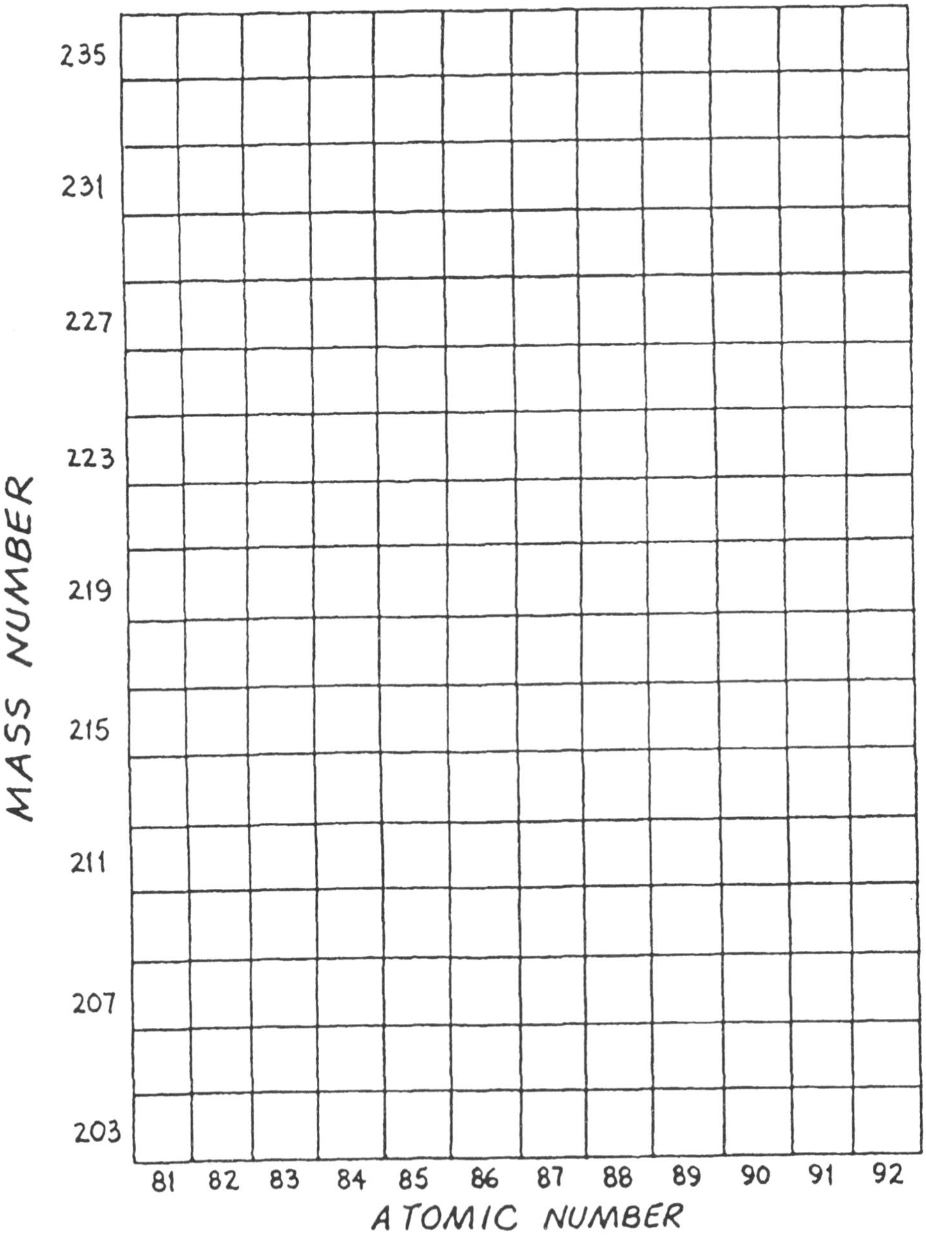

MASS NUMBER

235

231

227

223

219

215

211

207

203

81 82 83 84 85 86 87 88 89 90 91 92

A TOMIC NUMBER

What isotope is the final product? _____

CONCEPTUAL PHYSICAL SCIENCE **EXPLORATIONS**

Chapter 19 Radioactivity

Radioactive Half-Life

You and your classmates will now play the "half-life game." Each of you should have a coin to shake inside cupped hands. After it has been shaken for a few seconds, the coin is tossed on the table or on the floor. Students with tails up fall out of the game. Only those who consistently show heads remain in the game. Finally everybody has tossed a tail and the game is over.

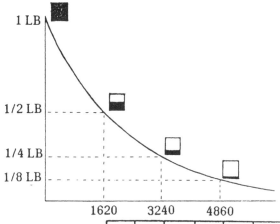

1. The graph to the left shows the decay of Radium-226 with time. Note that each 1620 years, half remains (the rest changes to other elements). In the grid below, plot the number of students left in the game after each toss. Draw a smooth curve that passes close to the points on your plot. What is the similarity of your curve with that of the curve of Radium-226?

_____

_____

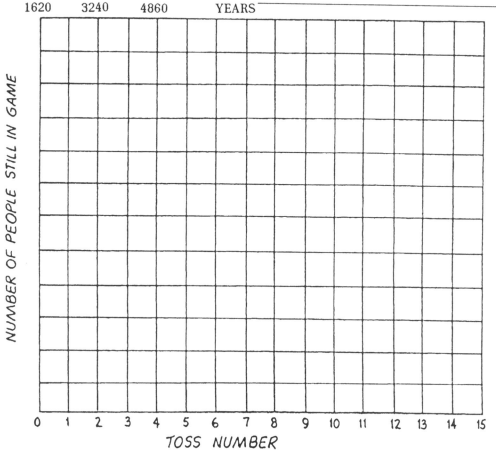

2. Was the person to last longest in the game *lucky*, with some sort of special powers to guide the long survival? What test could you make to decide the answer to this question?

_____

Chapter 20 Nuclear Fission and Fusion

Nuclear Fission and Fusion

1. Complete the table for a chain reaction in which two neutrons from each step individually cause a new reaction.

EVENT	1	2	3	4	5	6	7
NO. OF REACTIONS	1	2	4				

2. Complete the table for a chain reaction where three neutrons from each reaction cause a new reaction.

EVENT	1	2	3	4	5	6	7
NO. OF REACTIONS	1	3	9				

3. Complete these beta reactions, which occur in a breeder reactor.

$$^{239}_{92}U \longrightarrow \underline{\hspace{1.5cm}} + ^{0}_{-1}e$$

$$^{239}_{93}Np \longrightarrow \underline{\hspace{1.5cm}} + ^{0}_{-1}e$$

4. Complete the following fission reactions.

$$^{1}_{0}n + ^{235}_{92}U \longrightarrow ^{143}_{54}Xe + ^{90}_{38}Sr + \underline{\hspace{1cm}} \left(^{1}_{0}n\right)$$

$$^{1}_{0}n + ^{235}_{92}U \longrightarrow ^{152}_{60}Nd + \underline{\hspace{1cm}} + 4\left(^{1}_{0}n\right)$$

$$^{1}_{0}n + ^{239}_{94}Pu \longrightarrow \underline{\hspace{1cm}} + ^{97}_{40}Zr + 2\left(^{1}_{0}n\right)$$

5. Complete the following fusion reactions.

$$^{2}_{1}H + ^{2}_{1}H \longrightarrow ^{3}_{2}He + \underline{\hspace{1cm}}$$

$$^{2}_{1}H + ^{3}_{1}H \longrightarrow ^{4}_{2}He + \underline{\hspace{1cm}}$$

KNOW NUKES!

Nuclear Reactions

Complete these nuclear reactions.

1. $^{238}_{92}\text{U} \longrightarrow \, ^{234}_{90}\text{Th} + \, ^{4}_{2}\underline{\quad}$

2. $^{234}_{90}\text{Th} \longrightarrow \, ^{234}_{91}\text{Pa} + \, ^{0}_{-1}\underline{\quad}$

3. $^{234}_{91}\text{Pa} \longrightarrow \underline{\quad} + \, ^{4}_{2}\text{He}$

4. $^{220}_{86}\text{Rn} \longrightarrow \underline{\quad} + \, ^{4}_{2}\text{He}$

5. $^{216}_{84}\text{Po} \longrightarrow \underline{\quad} + \, ^{0}_{-1}\text{e}$

6. $^{216}_{84}\text{Po} \longrightarrow \underline{\quad} + \, ^{4}_{2}\text{He}$

7. $^{210}_{83}\text{Bi} \longrightarrow \underline{\quad} + \, ^{0}_{-1}\text{e}$

8. $^{1}_{0}\text{n} + \, ^{10}_{5}\text{B} \longrightarrow \underline{\quad} + \, ^{4}_{2}\text{He}$

CONCEPTUAL PHYSICAL SCIENCE **EXPLORATIONS**

Chapter 21 Elements of Chemistry
The Submicroscopic

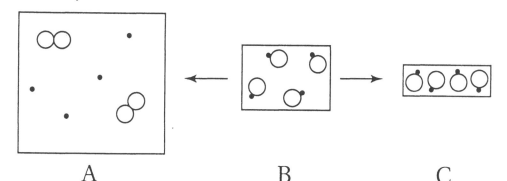

A   B   C

1. How many molecules are shown in   A _____   B _____   C _____
2. How many atoms are shown in   A _____   B _____   C _____
3. Which represents a physical change?   B ➝ A   B ➝ C   (circle one)
4. Which represents a chemical change?   B ➝ A   B ➝ C   (circle one)
5. Which box(es) represent(s) a mixture?   A _____   B _____   C _____
6. Which box contains the most mass?   A _____   B _____   C _____
7. Which box is coldest?   A _____   B _____   C _____
8. Which box contains the most air
   between molecules?   A _____   B _____   C _____

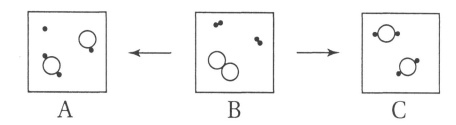

A   B   C

9. How many molecules are shown in   A _____   B _____   C _____
10. How many atoms are shown in   A _____   B _____   C _____
11. Which represents a physical change?   B ➝ A   B ➝ C   (circle one)
12. Which represents a chemical change?   B ➝ A   B ➝ C   (circle one)
13. Which box(es) represent(s) a mixture? A _____   B _____   C _____
14. Which box contains the most mass?   A _____   B _____   C _____
15. Which should take longer?   B ➝ A   B ➝ C   (circle one)
16. Which box most likely contains ions?   A _____   B _____   C _____

CONCEPTUAL PHYSICAL SCIENCE **EXPLORATIONS**

Chapter 21 Elements of Chemistry
Balancing Chemical Equations

In a balanced chemical equation the number of times each element appears as a reactant is equal to the number of times it appears as a product. For example,

$$2\ H_2\ +\ O_2\ \text{--->}\ 2\ H_2O$$

Recall that *coefficients* (the integer appearing before the chemical formula) indicate the number of times each chemical formula is to be counted and *subscripts* indicate when a particular element occurs more than once within the formula.

Check whether or not the following chemical equations are balanced.

$$3\ NO\ \text{--->}\ N_2O\ +\ NO_2$$ ☐ balanced; ☐ unbalanced

$$SiO_2\ +\ 4\ HF\ \text{--->}\ SiF_4\ +\ 2\ H_2O$$ ☐ balanced; ☐ unbalanced

$$4\ NH_3\ +\ 5\ O_2\ \text{--->}\ 4\ NO\ +\ 6\ H_2O$$ ☐ balanced; ☐ unbalanced

Unbalanced equations are balanced by changing the coefficients. Subscripts, however, should never be changed because this changes the chemical's identity—H_2O is water, but H_2O_2 is hydrogen peroxide! The following steps may help guide you:

1. Focus on balancing only one element at a time. Start with the left-most element and modify the coefficients such that this element appears on both sides of the arrow the same number of times.

2. Move to the next element and modify the coefficients so as to balance this element. Do not worry if you incidently unbalance the previous element. You will come back to it in subsequent steps.

3. Continue from left to right balancing each element individually.

4. Repeat steps 1 - 3 until all elements are balanced.

Use the above methodology to balance the following chemical equations.

____N_2O +   ____N_2 --->   ____O_2

____$NaClO_3$ --->   ____$NaCl$ +   ____O_2

____$MnCl_2$ +   ____Al --->   ____Mn +   ____$AlCl_3$

____K +   ____H_2O --->   ____H_2 +   ____KOH

____Al_2O_3 +   ____C --->   ____Al +   ____CO_2

____NH_3 +   ____F_2 --->   ____NH_4F +   ____NF_3

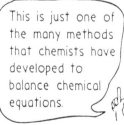

This is just one of the many methods that chemists have developed to balance chemical equations.

Knowing how to balance a chemical equation is a useful technique, but understanding why a chemical equation needs to be balanced in the first place is far more important.

Chapter 21 Elements of Chemistry
Physical and Chemical Changes

1. What distinguishes a chemical change from a physical change?

2. Based upon observations alone, why is distinguishing a chemical change from a physical change not always so straight-forward?

Try your hand at catagorizing the following processes as either chemical or physical changes. Some of these examples are debatable! Be sure to discuss your reasoning with fellow classmates or your instructor.

(circle one)

3. A cloud grows dark - chemical   physical

4. Leaves produce oxygen. - chemical   physical

5. Food coloring is added to water. - - - - - - - - - - - - - - chemical   physical

6. Tropical coral reef dies. - chemical   physical

7. Dead coral reef is pounded by waves into beach sand. - - - - chemical   physical

8. Oil and vinegar separate. - chemical   physical

9. Soda drink goes flat. - chemical   physical

10. Sick person develops a fever. - - - - - - - - - - - - - - - - - chemical   physical

11. Compost pit turns into mulch - - - - - - - - - - - - - - - - chemical   physical

12. A computer is turned on. - - - - - - - - - - - - - - - - - - - chemical   physical

13. An electrical short melts a computer's integrated circuits. - chemical   physical

14. A car battery runs down. - chemical   physical

15. A pencil is sharpened. - chemical   physical

16. Mascara is applied to eyelashes. - - - - - - - - - - - - - - - chemical   physical

17. Sunbather gets tan lying in the sun. - - - - - - - - - - - - - chemical   physical

18. Invisible ink turns visible upon heating. - - - - - - - - - - chemical   physical

19. A light bulb burns out. - chemical   physical

20. Car engine consumes a tank of gasoline. - - - - - - - - - chemical   physical

21. B vitamins turn urine yellow. - - - - - - - - - - - - - - - - - chemical   physical

CONCEPTUAL PHYSICAL SCIENCE EXPLORATIONS
Chapter 22 Mixtures
Pure Mathematics

Using a scientist's definition of pure, identify whether each of the following is 100% pure:

	100% pure?	
Freshly squeezed orange juice	Yes	No
Country air	Yes	No
Ocean water	Yes	No
Fresh drinking water	Yes	No
Skim milk	Yes	No
Stainless steel	Yes	No
A single water molecule	Yes	No

A glass of water contains on the order of a trillion trillion (1×10^{24}) molecules. If the water in this were 99.9999% pure, you could calculate the percent of impurities by subtracting from 100.0000%

$$\begin{array}{r} 100.0000\% \text{ water } + \text{ impurity molecules} \\ - 99.9999\% \text{ water molecules} \\ \hline 0.0001\% \text{ impurity molecules} \end{array}$$

Pull out your calculator and calculate the number of impurity molecules in the glass of water. Do this by finding 0.0001% of 1×10^{24}, which is the same as muliplying 1×10^{24} by 0.000001.

$$(1 \times 10^{24})(0.000001) = \underline{\hspace{4cm}}$$

How many impurity molecules are there in a glass of water that's 99.9999% pure?

a) 1,000 (one thousand: 10^3) b) 1,000,000 (one million: 10^6)

c) 1,000,000,000 (one billion: 10^9) d) 1,000,000,000,000,000,000 (one million trillion: 10^{18}).

How does your answer make you feel about drinking water that is 99.9999 percent free of some poison, such as a pesticide?

_____

For every one impurity molecule, how many water molecules are there? (Divide the number of water molecules by the number of impurity molecules.)

_____

Would you describe these impurity molecules within water that's 99.9999% pure as "rare" or "common"?

_____

A friend argues that he or she doesn't drink tap water because it contains thousands of molecules of some impurity in each glass. How would you respond in defense of the water's purity, if it indeed does contain thousands of molecules of some impurity per glass?

_____

CONCEPTUAL PHYSICAL SCIENCE **EXPLORATIONS**

Chapter 23 Chemical Bonding
Chemical Bonds

1. Based upon their positions in the periodic table, predict whether each pair of elements will form an ionic, covalent, or neither (atomic number in parenthesis)

    a. Gold (79) and Platinum (78) _____   b. Rubidium (37) and Iodine (53) _____

    c. Sulfur (16) and Chlorine (17) _____   d. Sulfur (16) and Magnesium (12) _____

    e. Calcium (20) and Chlorine (17) _____   f. Germanium(32) and Arsenic (33) _____

    g. Iron (26) and Chromium (24) _____   h. Chlorine (17) and Iodine (53) _____

    i. Carbon (6) and Bromine (35) _____   j. Barium (56) and Astatine (85) _____

2. The most common ions of lithium, magnesium, aluminum, chlorine, oxygen, and nitrogen nitrogen and their respective charges are as follows:

Positively Charged Ions	Negatively Charged Ions
Lithium ion: Li^{1+}	Chloride ion: Cl^{1-}
Barium ion: Ba^{2+}	Oxide ion: O^{2-}
Aluminum ion: Al^{3+}	Nitride ion: N^{3-}

Use this information to predict the chemical formulas for the following ionic compounds:

    a. Lithium Chloride:_____   b. Barium Chloride: _____   c. Aluminum Chloride: _____

    d. Lithium Oxide:_____   e. Barium Oxide: _____   f. Aluminum Oxide: _____

    g. Lithium Nitride:_____   h. Barium Nitride: _____   i. Aluminum Nitride: _____

    j. How are elements that form positive ions grouped in the periodic table relative to

       elements that form negative ions?_____

3. Predict whether the following chemical structures are polar or nonpolar:

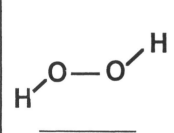

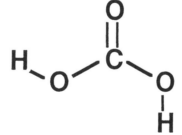

CONCEPTUAL PHYSICAL SCIENCE **EXPLORATIONS**

Chapter 23 Chemical Bonding
Shells and the Covalent Bond

When atoms bond covalently their atomic shells overlap so that shared electrons can occupy both shells at the same time.

Non-bonded hydrogen atoms

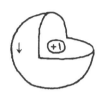

Hydrogen   Hydrogen

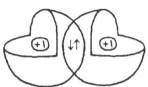

Covalently bonded hydrogen atoms

Molecular Hydroger

Formula: H_2

Fill each shell model shown below with enough electrons to make each atom electrically neutral. Use arrows to represent electrons. Within the box draw a sketch showing how the two atoms bond covalently. Draw hydrogen shells more than once when necessary so that no electrons remain unpaired. Write the name and chemical formula for each compound.

A.

Hydrogen   Carbon

Name of Compound: Formula:

B.

Hydrogen   Nitroger

Name of Compound: Formula:

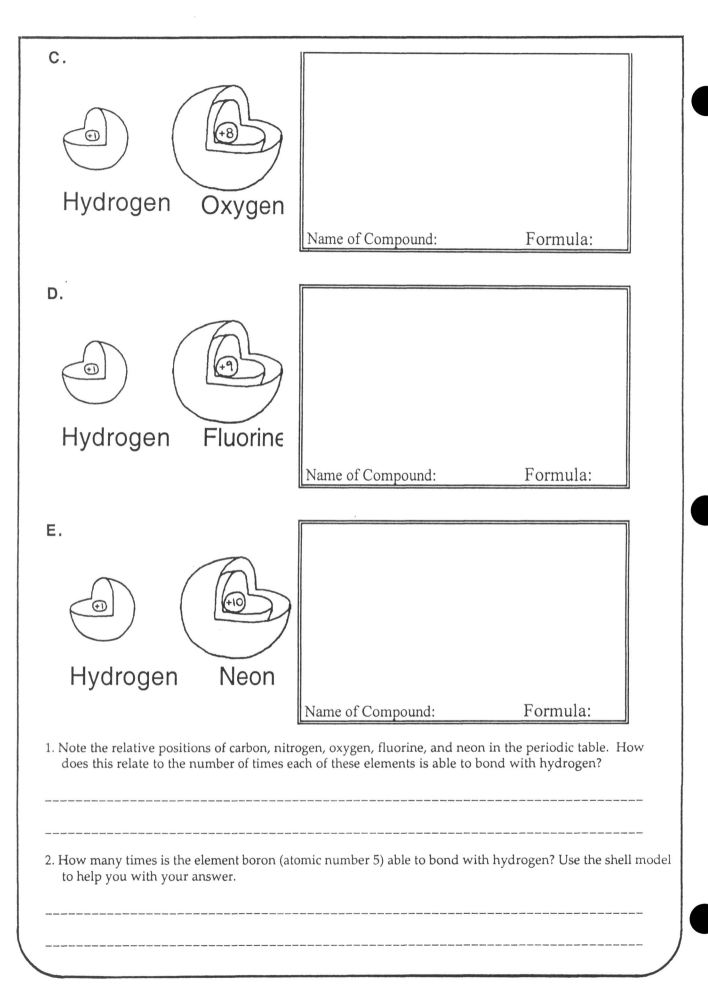

C.

Hydrogen   Oxygen

Name of Compound: Formula:

D.

Hydrogen   Fluorine

Name of Compound: Formula:

E.

Hydrogen   Neon

Name of Compound: Formula:

1. Note the relative positions of carbon, nitrogen, oxygen, fluorine, and neon in the periodic table. How does this relate to the number of times each of these elements is able to bond with hydrogen?

_____

_____

2. How many times is the element boron (atomic number 5) able to bond with hydrogen? Use the shell model to help you with your answer.

_____

_____

CONCEPTUAL PHYSICAL SCIENCE EXPLORATIONS

Chapter 23 Chemical Bonding
Bond Polarity and the Shell Model

Pretend you are one of two electrons being shared by a hydrogen atom and a fluorine atom. Say, for the moment, you are centrally located between the two nuclei. You find that both nuclei are attracted to you, hence, because of your presence the two nuclei are held together.

You are here

H : F

1. Why are the nuclei of these atoms attracted to you?_____

2. What type of chemical bonding is this?_____

You are held within hydrogen's 1st shell and at the same time within fluorine's 2nd shell. Draw a sketch using the shell models below to show how this is possible. Represent yourself and all other electrons usings arrows. Note your particular location.

Hydrogen Fluorine

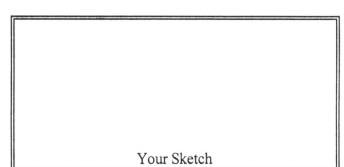

Your Sketch

According to the laws of physics, if the nuclei are both attracted to you, then you are attracted to both of the nuclei.

3. You are pulled toward the hydrogen nucleus, which has a positive charge. How strong is this charge from your point of view—what is its *electronegativity?* _____

4. You are also attracted to the fluorine nucleus. What is its electronegativity?____

 You are being shared by the hydrogen and fluorine nuclei. But as a moving electron you have some choice as to your location.

5. Consider the electronegativities you experience from both nuclei. Which nucleus would you tend to be closest to? _____

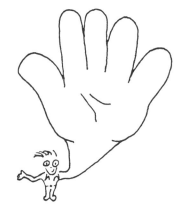

Stop pretending you are an electron and observe the hydrogen-fluorine bond from outside the hydrogen fluoride molecule. Bonding electrons tend to congregate to one side because of the differences in effective nuclear charges. This makes one side slightly negative in character and the opposite side slightly positive. Indicate this on the following structure for hydrogen fluoride using the symbols δ- and δ+

$$H \;\; \overset{..}{} \;\; F$$

By convention, bonding electrons are not shown. Instead, a line is simply drawn connecting the two bonded atoms. Again indicate the slightly negative and positive ends.

$$H - F$$

6. Would you describe hydrogen fluoride as a polar or nonpolar molecule?_____

7. If two hydrogen fluoride molecules were thrown together would they stick or repel? (Hint: what happens when you throw two small magnets together?)_____

8. Place bonds between the hydrogen and fluorine atoms to show many hydrogen fluoride molecules grouped together. Each element should be bonded only once. Circle each molecule and indicate the slightly negative and slightly positive ends.

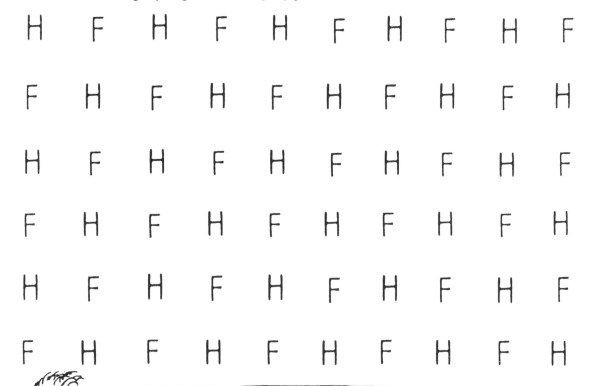

The interactions that occur between molecules is the subject of Chapter 24 in your textbook. Onward !

CONCEPTUAL PHYSICAL SCIENCE EXPLORATIONS

Chapter 24 Molecular Mixing
Atoms to Molecules to Molecular Attractions

protons   neutrons   electrons

SUBATOMIC PARTICLES

Subatomic particles are the fundamental building blocks of all _____

hydrogen atom

hydrogen atom

oxygen atom

An atom is a group of _____ held tightly together. An oxygen atom is a group of 8 _____ , 8 _____, and 8 _____. A hydrogen atom is a group of only 1 _____ and 1 _____

oxygen atom

hydrogen atom

hydrogen atom

ATOMS

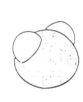

water molecule

water molecule

A _____ is a group of atoms held tightly together. A water _____ consists of 2 _____ atoms and 1 _____ atom.

MOLECULES

WATER

Water is a material made up of billions upon billions of water _____ . The physical properties of water are based upon how these water _____ interact with one another. The electronic attractions between _____ is the main topic of Chapter 24.

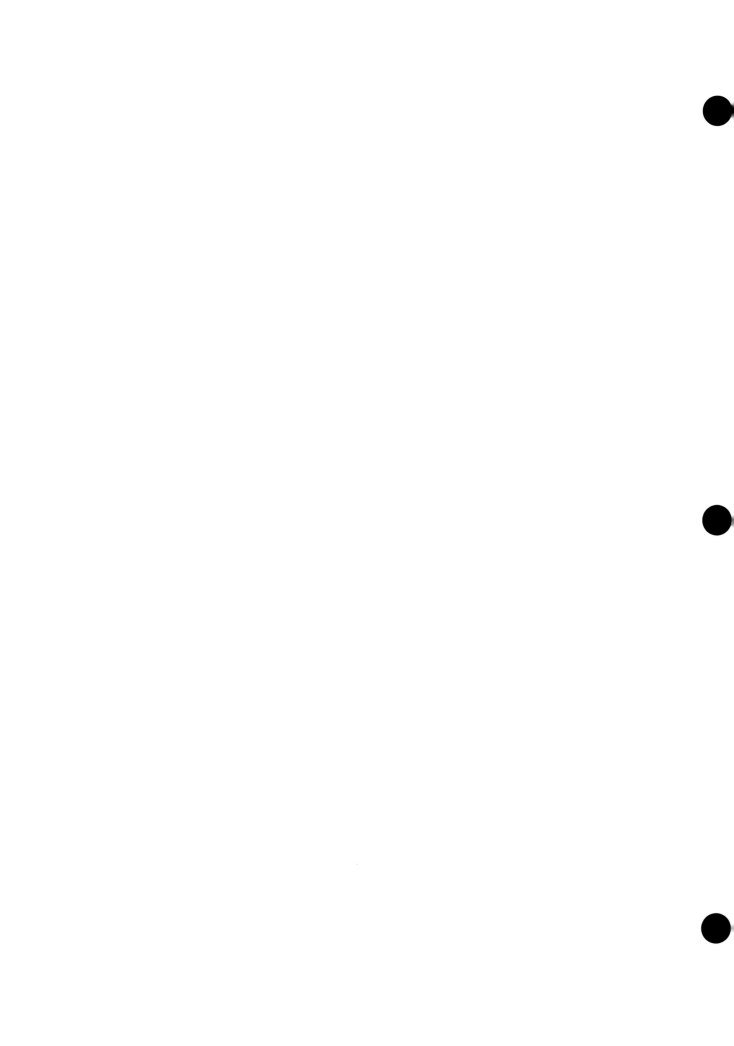

CONCEPTUAL PHYSICAL SCIENCE **EXPLORATIONS**

Chapter 24 Molecular Mixing
Solutions

1. Use these terms to complete the following sentences. Some terms may be used more than once.

solution	solvent	solute
dissolve	concentrated	dilute
saturated	concentration	mole
molarity	solubility	soluble
insoluble	precipitate	supersaturated

Sugar is _____ in water for the two can be mixed homogeneously to form a _____.

The _____ of sugar in water is so great that _____ homogeneous mixtures are easily

prepared. Sugar, however, is not infinitely _____ in water for when too much of this

_____ is added to water, which behaves as the _____, the solution becomes

_____. At this point any additional sugar is _____ for it will not _____. If

the temperature of a saturated sugar solution is lowered, the _____ of the sugar in water is also

lowered. If some of the sugar comes out of solution, it is said to form a _____. If, however, the

sugar remains in solution despite the decrease in solubility, then the solution is said to be

_____. Adding only a small amount of sugar to water results in a _____ solution.

The _____ of this solution or any solution can be measure in terms of _____, which

tells us the number of solute molecules per liter of solution. If there are 6.022×10^{23} molecules in 1 liter

of solution, then the _____ of the solution is 1 _____ per liter.

2. Temperature has a variety of effects on the solubilites of various solutes. With some solutes, such as sugar, solubility increases with increasing temperature. With other solutes, such as sodium chloride (table salt), changing temperature has no significant effect. With some solutes, such as lithium sulfate, Li_2SO_4, the solubility actually decreases with increasing temperature.

a. Describe how you would prepare a supersaturated solution of lithium sulfate.

_____

b. How might you cause a saturated solution of lithium sulfate to form a precipitate?

_____

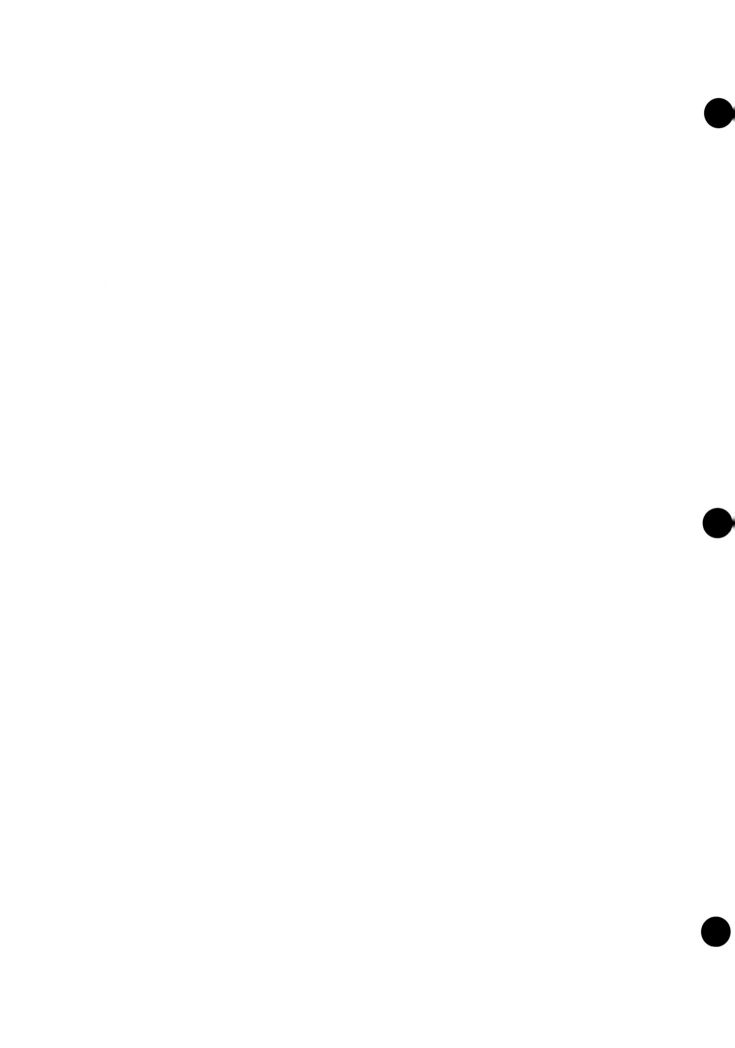

CONCEPTUAL PHYSICAL SCIENCE **EXPLORATIONS**

Chapter 25 Acids and Bases
Donating and Accepting Hydrogen Ions

A chemical reaction that involves the transfer of a hydrogen ion from one molecule to another is classified as an acid-base reaction. The molecule that donates the hydrogen ion behaves as an acid. The molecule that accepts the hydrogen ion behaves as a base.

On paper, the acid-base process can be depicted through a series of frames:

frame 1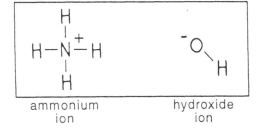

Ammonium and hydroxide ions in close proximity.

frame 2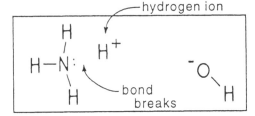

Bond is broken between the nitrogen and a hydrogen of the ammonium ion. The two electrons of the broken bond stay with the nitrogen leaving the hydrogen with a positive charge.

frame 3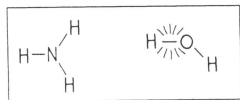

The hydrogen ion migrates to the hydroxide ion.

frame 4

The hydrogen ion bonds with the hydroxide ion to form a water molecule.

In equation form we abbreviate this process by only showing the before and after:

Donating and Accepting Hydrogen Ions continued:

We see from the previous reaction that because the ammonium ion donated a hydrogen ion, it behaved as an acid. Conversely, the hydroxide ion by accepting a hydrogen ion behaved as a base. How do the ammonia and water molecules behave during the reverse process?

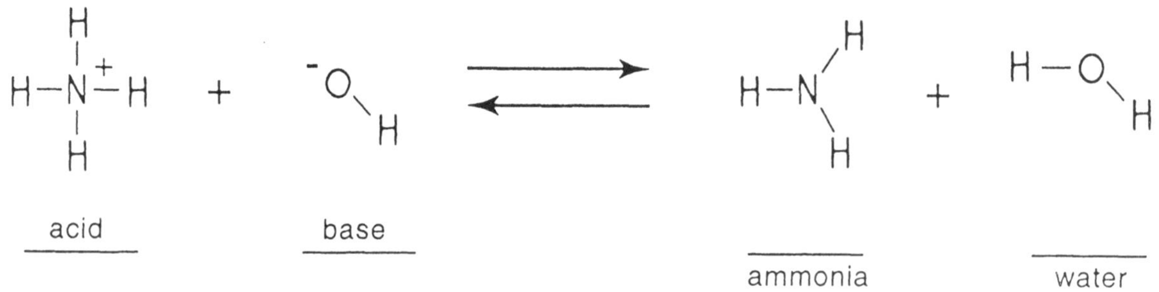

$$H-\overset{\overset{\displaystyle H}{|}}{\underset{\underset{\displaystyle H}{|}}{N}}\overset{+}{-}H \quad + \quad {}^{-}\overset{\displaystyle O}{\diagdown}_{H} \quad \rightleftharpoons \quad H-\overset{\overset{\displaystyle H}{}}{\underset{\underset{\displaystyle H}{}}{N}} \quad + \quad H-\overset{\displaystyle O}{\diagdown}_{H}$$

    acid   base   ammonia   water

Identify the following molecules as behaving as an acid or a base:

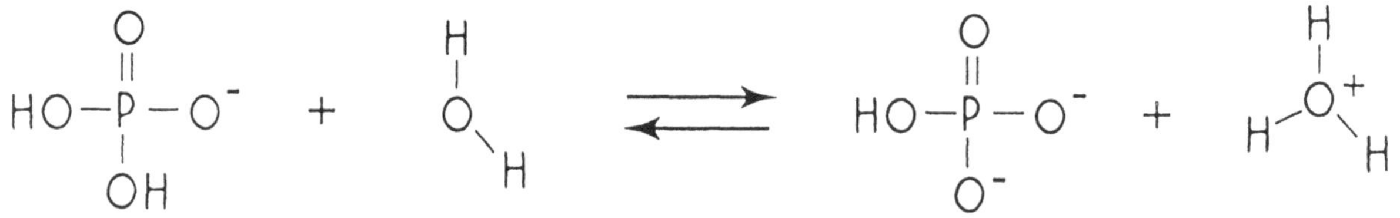

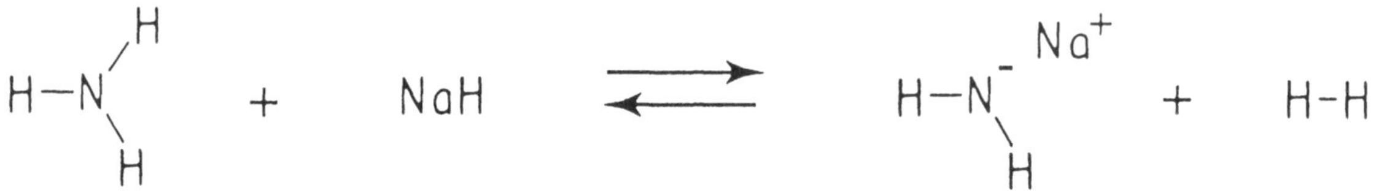

$$HO-\overset{\overset{\displaystyle O}{||}}{\underset{\underset{\displaystyle OH}{|}}{P}}-O^{-} \quad + \quad \overset{\displaystyle H}{\underset{\underset{\displaystyle H}{|}}{O}} \quad \rightleftharpoons \quad HO-\overset{\overset{\displaystyle O}{||}}{\underset{\underset{\displaystyle O^{-}}{|}}{P}}-O^{-} \quad + \quad H-\overset{\overset{\displaystyle H}{|}}{\underset{\underset{\displaystyle H}{}}{O}}{}^{+}$$

_____   _____   _____   _____

$$H-\overset{\overset{\displaystyle H}{}}{\underset{\underset{\displaystyle H}{}}{N}} \quad + \quad NaH \quad \rightleftharpoons \quad H-\overset{\overset{\displaystyle {}^{-}\,Na^{+}}{}}{\underset{\underset{\displaystyle H}{}}{N}} \quad + \quad H-H$$

_____   _____   _____   _____

$$H-H \quad + \quad {}^{-}H \quad \rightleftharpoons \quad H^{-} \quad + \quad H-H$$

_____   _____   _____   _____

$$HNO_3 \quad + \quad NH_3 \quad \rightleftharpoons \quad {}^{-}NO_3 \quad + \quad {}^{+}NH_4$$

_____   _____   _____   _____

CONCEPTUAL PHYSICAL SCIENCE **EXPLORATIONS**
Chapter 26 Oxidation and Reduction
Loss and Gain of Electrons

A chemical reaction that involves the transfer of an electron is classified as an oxidation-reduction reaction. Oxidation is the process of losing an electrons, while reduction is the process of gaining them. Any chemical that causes another chemical to lose electrons (become oxidized) is called an oxidizing agent. Conversely, any chemical that causes another chemical to gain electrons is called a reducing agent.

1. What is the relationship between an atom's ability to behave as an oxidizing agent and its electron affinity?

2. Relative to the periodic table, which elements tend to behave as strong oxidizing agents?

3. Why don't the noble gases behave as oxidizing agents?

4. How is it that an oxidizing agent is itself reduced?

5. Is it possible to have an endothermic oxidation-reduction reaction? If so, cite examples.

6. Specify whether each reactant is about to be oxidized or reduced.

$$2\ K \quad + \quad H_2O \quad \longrightarrow \quad 2\ K^+ \quad + \quad {}^-OH$$
_____   _____

$$2\ Mg \quad + \quad O_2 \quad \longrightarrow \quad 2\ Mg^{2+}O^{2-}$$
_____   _____

$$2\ Na \quad + \quad Cl_2 \quad \longrightarrow \quad 2\ Na^+Cl^-$$
_____   _____

$$CH_4 \quad + \quad 2\ O_2 \quad \longrightarrow \quad O{=}C{=}O \quad + \quad {}_H^{\;\;O-H}$$
_____   _____

7. Which oxygen atom enjoys a greater negative charge?

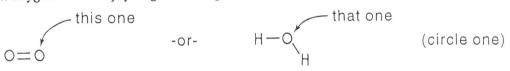

this one   that one

$O{=}O$ -or- $H-O\underset{H}{\diagdown}$ (circle one)

8. Relate your answer to question 7 to how it is that O_2 is reduced upon reacting with CH_4 to form carbon dioxide and water.

CONCEPTUAL PHYSICAL SCIENCE **EXPLORATIONS**

Chapter 27 Organic Compounds
Structures of Organic Compounds

1. What are the chemical formulas for the following structures.

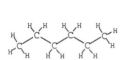 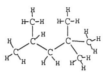 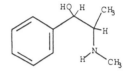

Formula: _____   _____   _____   _____

2. How many covalent bonds is carbon able to form?_____

3. What is wrong with the structure shown in the box:

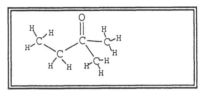

_____

4a. Draw a hydrocarbon that contains 4 carbon atoms	4b. Redraw your structure and transform it into an amine.	4c. Transform your amine into an amide. You may need to relocate the nitrogen.

4d. Redraw your amide transforming it into a carboxylic acid.	4e. Redraw your carboxylic acid transforming it into an alcohol.	4f. Rearrange the carbons of your alcohol to make an ether.

5. Circle the following alkaloids that are in their free-base form?

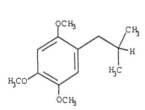 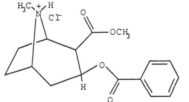 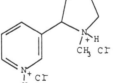 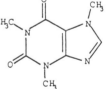

     Mescaline   Cocaine   Nicotine   Caffeine

6. How might you convert a free-base alkaloid into a salt?

7. Why are alkaloids less water soluble in their free-base form?

8. Which should have a greater tendency to vaporize upon heating: an alkaloid in its free-base form or one in its salt from? Why?

CONCEPTUAL PHYSICAL SCIENCE **EXPLORATIONS**

Chapter 28 Chemistry of Drugs
Neurotransmitters

Many drugs work by mimicking neurotransmitters. Amphetamine, for example, mimicks the stimulatory neurotransmitter norepinephrine. Much of this occurs in the synaptic cleft, which is a narrow gap between neurons.

Stick
Structures:

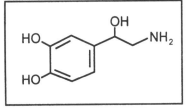

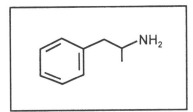

Norepinephrine   Amphetamine

Schematic
Representation:

1. Under normal conditions, the pre-synaptic neuron releases norepinephrine, which migrates across the synaptic cleft to a receptor site on the surface of the post-synaptic neuron.

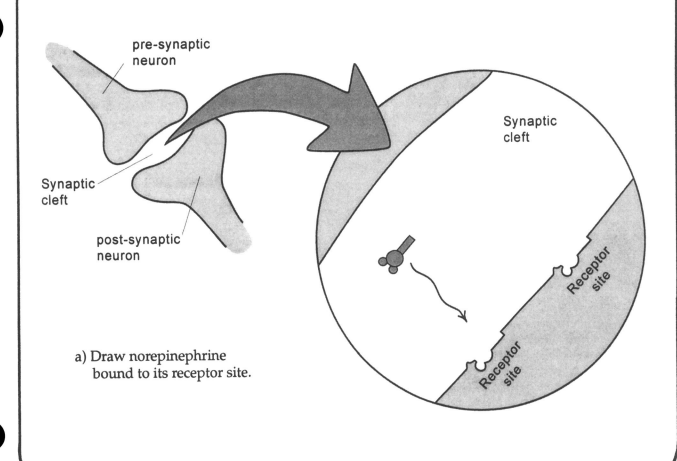

a) Draw norepinephrine bound to its receptor site.

2. When amphetamine molecules get within the cleft, there is too much stimulation of the post-synaptic neuron.

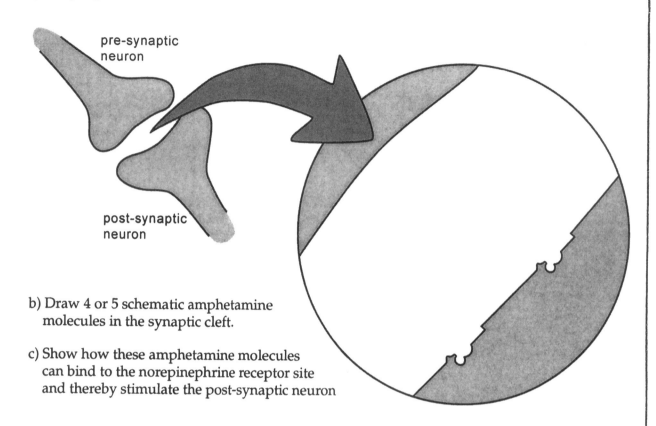

b) Draw 4 or 5 schematic amphetamine molecules in the synaptic cleft.

c) Show how these amphetamine molecules can bind to the norepinephrine receptor site and thereby stimulate the post-synaptic neuron

3. In response to the overstimulation, the body stops producing so much norepinephrine. Also, the amphetamine molecules are eventually broken down and the result is shown below:

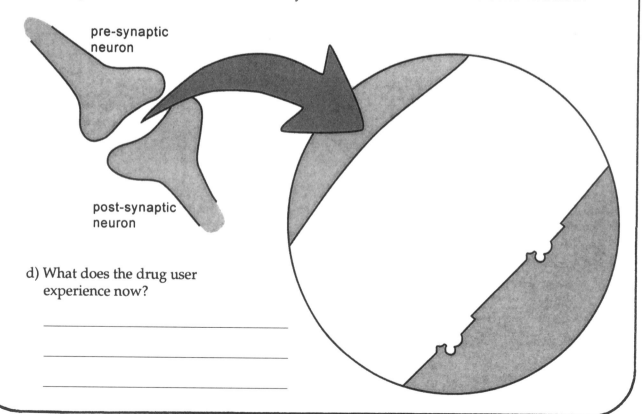

d) What does the drug user experience now?

_____

_____

_____

CONCEPTUAL PHYSICAL SCIENCE EXPLORATIONS

Chapter 29 Plastics
Polymers

1. Circle the monomers that may be useful for forming an addition polymer and draw a
   box around the ones that may be useful for forming a condensation polymer.

2. Which type of polymer always weighs less than the sum of its parts? Why?

3. Would a material with the following arrangement of polymer molecules have a relatively
   high or low melting point? Why?

105

Chapter 29 Plastics
History of Plastics

Turn the page sideways. Read the descriptions of plastics and using Sections 29.3 – 29.6 of your textbook, identify the name of each plastic for that time. Write its name in the blank that shows its correct time of its discovery.

For example, write **rubber vulcanization** in the blank for 1839.

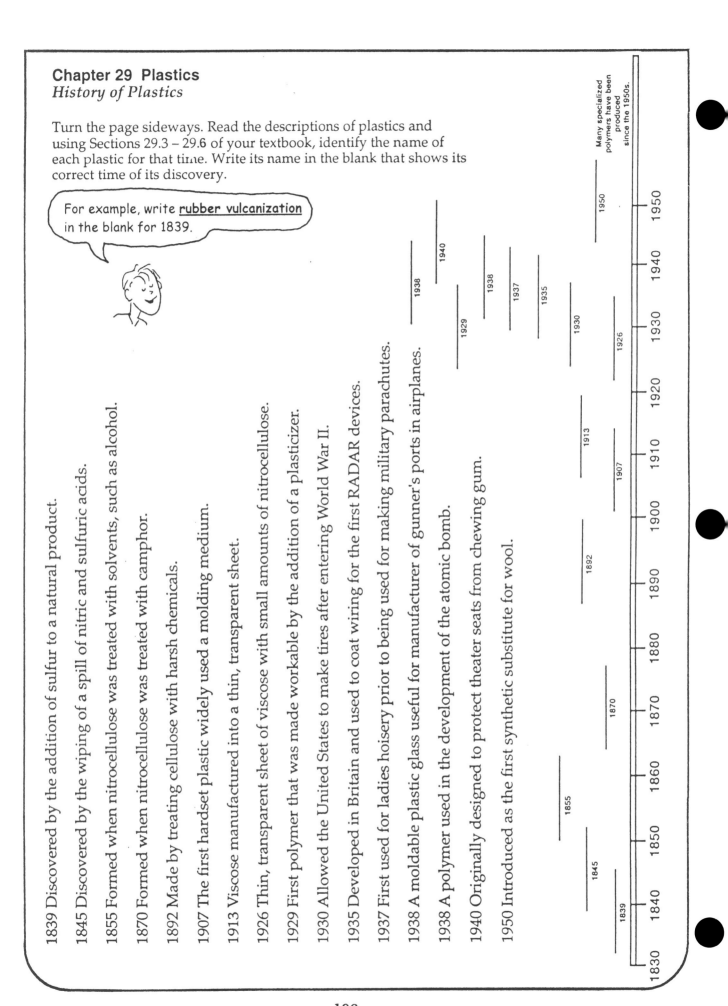

1839 Discovered by the addition of sulfur to a natural product.

1845 Discovered by the wiping of a spill of nitric and sulfuric acids.

1855 Formed when nitrocellulose was treated with solvents, such as alcohol.

1870 Formed when nitrocellulose was treated with camphor.

1892 Made by treating cellulose with harsh chemicals.

1907 The first hardset plastic widely used a molding medium.

1913 Viscose manufactured into a thin, transparent sheet.

1926 Thin, transparent sheet of viscose with small amounts of nitrocellulose.

1929 First polymer that was made workable by the addition of a plasticizer.

1930 Allowed the United States to make tires after entering World War II.

1935 Developed in Britain and used to coat wiring for the first RADAR devices.

1937 First used for ladies hoisery prior to being used for making military parachutes.

1938 A moldable plastic glass useful for manufacturer of gunner's ports in airplanes.

1938 A polymer used in the development of the atomic bomb.

1940 Originally designed to protect theater seats from chewing gum.

1950 Introduced as the first synthetic substitute for wool.

Many specialized polymers have been produced since the 1950s.

Name   Period   Date

CONCEPTUAL PHYSICAL SCIENCE **EXPLORATIONS**

Chapter 30   Minerals and Their Formation
Chemical Structure and Formulas of Minerals

Out of the more than 3400 minerals, only about two dozen are abundant. Minerals are classified by their chemical composition and internal atomic structure and are divided into groups. For this exercise we explore the following mineral groups: *carbonates, sulfides, sulfates,* and *halides.*

For each mineral structure diagrammed below, look for a pattern in the structure, count the number of atoms (ions) in each, and fill in the blanks.

The schematic diagrams are simple representations of small mineral structures. Actual mineral structures extend farther and comprise more atoms.

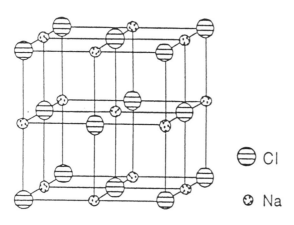

⊖ Cl

⊘ Na

1. Circle pairs of Na and Cl ions in the structure and add any ion(s) needed to complete pairing. This mineral structure contains _____ Na ions and _____ Cl ions. The mineral's formula is _____. This mineral belongs to the _____ group.

2. This mineral structure contains _____ Ca atoms, ____ C atoms, and _____ O atoms. The mineral's formula is _____. This mineral belongs to the _____ group.

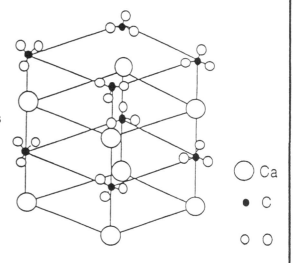

○ Ca

• C

○ O

107

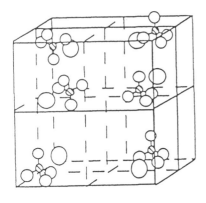

3. This mineral structure contains _____ Ca atoms, _____ S atoms, and _____ O ions. The mineral's formula is _____ This mineral belongs to the _____ group.

○ Ca

⊘ S

○ O

4. Complete the structure by adding the needed atom(s). This mineral structure contains _____ Fe atoms and _____ S atoms. The mineral's formula is _____. This mineral belongs to the _____ group.

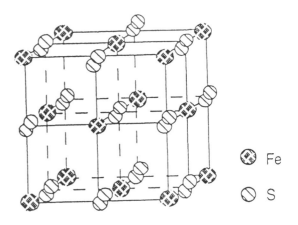

⊛ Fe

⊘ S

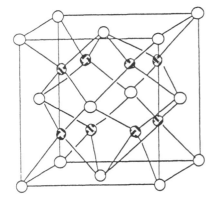

5. Complete the mineral structure so that each Ca atom is linked to two F atoms. Now the mineral structure contains _____ Ca atoms and _____ F atoms. The mineral's formula is _____. This mineral belongs to the _____ group.

○ Ca

⊘ F

CONCEPTUAL PHYSICAL SCIENCE EXPLORATIONS
Chapter 31   Rocks
The Rock Cycle

Complete the illustration, which depicts the different paths in the rock cycle. Insert arrows to show direction of pathways.

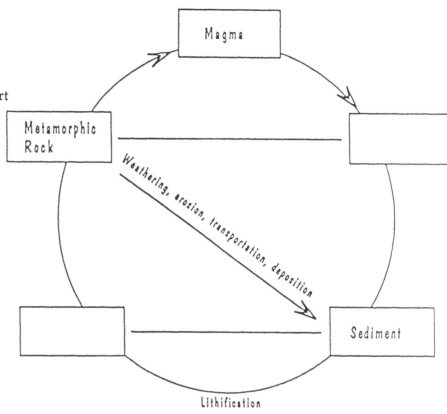

1. Can a rock that has undergone metamorphism turn into sedimentary rock? If so, how? If not, how come?

_____

2. By what process does hot molten magma become rock?

_____

3. List three rocks generated from the different types of magma.

_____

4. Carbonate rocks found in the Colorado Rocky Mountains imply what type of deposition environment?

_____

5. In what section(s) of the rock cycle are gemstones formed?

_____

6. The Big Island of Hawaii is almost 1 million years old. Yet hikers there almost never step on rock that is more than 1 thousand years old. Explain.

_____

Igneous Rock Differentiation: How to Make Granite

A mineral is called a high temperature mineral if its melting/freezing temperature is relatively high. A mineral is called a low temperature mineral if its melting/freezing temperature is relatively low.

Suppose we start with solid, basaltic rock. If it is heated, it will partially melt.

1. Is the type of mineral left behind (that doesn't liquefy) a high temperature or low temperature mineral?

_____

Do you think granite could form in this manner?

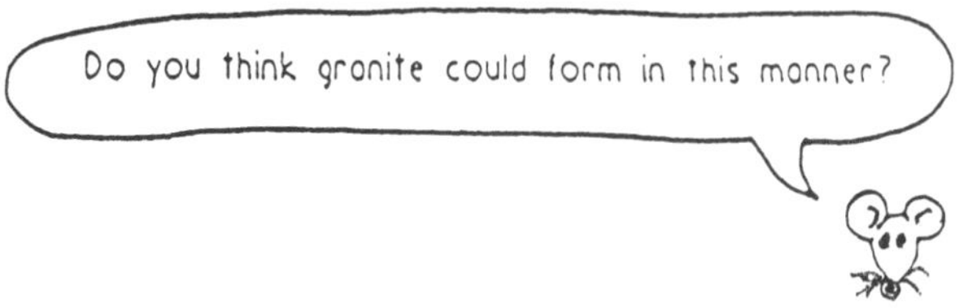

2. Which type will melt to form a liquid?

_____

3. Will the resulting liquid be higher or lower in silicon content than the original rock? Why?

_____

4. If this liquid is separated from the original rock and then cooled relatively quickly, what is the name of the rock that will most likely form?

_____

5. Repeat steps 1 through 4 for the rock formed in question 4. What is the name of the resulting rock if the liquid is allowed to cool very slowly?

_____

Now consider a magma chamber that contains completely molten basaltic magma. Let's allow this magma to cool very slowly.

6. Which type of minerals will be the first to form, low temperature or high temperature minerals?

_____

7. Will the remaining liquid be higher or lower in silicon content than the original liquid? Why?

_____

Assume that the newly formed crystals settle to the bottom of the magma chamber so that there is no chemical interaction between the newly formed crystals and the remaining liquid.

8. If this process continues, will the low temperature minerals eventually crystallize?

_____

9. If so, would a rock formed from these minerals be higher or lower in silicon content than a basalt?

_____

CONCEPTUAL PHYSICAL SCIENCE EXPLORATIONS

Chapter 32 The Architecture of the Earth
Faults

Three block diagrams are illustrated below. Draw arrows on each diagram to show the direction of movement. Answer the questions next to each diagram.

A.

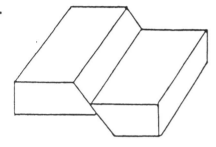

What type of force produced Fault A?

_____

Name the fault _____

Where would you expect to find this type of fault?

_____

_____

B.

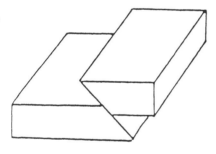

What type of force produced Fault B?

_____

Name the fault _____

Where would you expect to find this type of fault?

_____

_____

C.

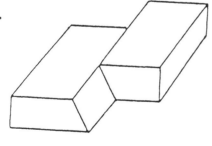

What type of force produced Fault C?

_____

Name the fault _____

Where would you expect to find this type of fault?

_____

_____

CONCEPTUAL PHYSICAL SCIENCE **EXPLORATIONS**
Chapter 32   The Architecture of the Earth
Structural Geology

Much subsurface information is learned by oil companies when wells are drilled. Some of this information leads to the discovery of oil, and some reveals subsurface structures such as folds and/or faults in the Earth's crust.

Four oil wells that have been drilled to the same depth are shown on the cross section below. Each well encounters contacts between different rock formations at the depths shown in the table below. Rock formations are labeled A — F, with A as the youngest rock formation and F as the oldest rock formation.

	Depth to Contact (in meters)			
Contact	Oil well #1	Oil well #2	Oil well #3	Oil well #4
A-B	200	not encountered	200	not encountered
B-C	400	100	400	100
C-D	600	300	600	300
D-E	800	500	800	500
E-F	1000	700	1000	700

1. In the cross section below, Contacts D - E and E – F are plotted for Oil Wells 1 and 2. Plot the remainder of the data for all four wells, labeling each point you plot.

2. Draw lines to connect the contacts between the rock formations (as is done for Contacts D - E and E – F for Oil Wells 1 and 2).

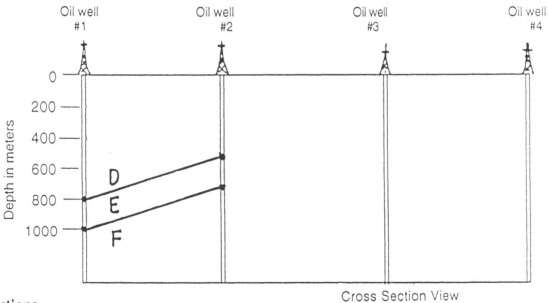

Cross Section View

Questions

1. What explanation can you offer for no sign of formation A in Wells 2 and 4?

_____

2. What geological structures are revealed? Label them on the cross section.

_____

CONCEPTUAL PHYSICAL SCIENCE **EXPLORATIONS**

Chapter 33 Our Restless Planet
Plate Boundaries

Draw arrows on the plate boundaries **A**, **B**, and **C**, to show the relative direction of movement.

Type of boundary for **A**? _____

What type of force generates this type of boundary?

A.

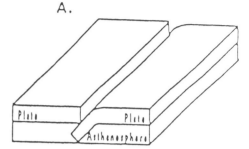

_____

Is this a site of lithospheric formation, destruction, or
lithospheric transport?

_____

Type of boundary for **B**? _____

B.

What type of force generates this type of boundary?

_____

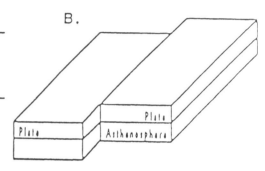

Is this a site of lithospheric formation, destruction, or
lithospheric transport?

_____

C.

Type of boundary for **C**? _____

What type of force generates this type of boundary?

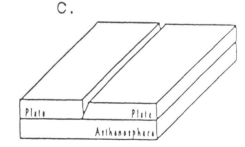

_____

Is this a site of lithospheric formation, destruction, or
lithospheric transport?

_____

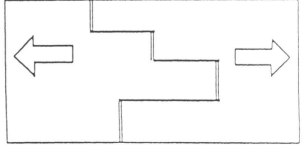

Draw arrows on the transform faults to

indicate relative motion.

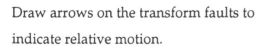

    ——— Transform fault

    ‖ Mid-ocean spreading ridge

Chapter 33 Our Restless Planet
Sea-Floor Spreading

The rate of sea floor spreading, some 1 to 10 centimeters per year, is found by knowing the distance and age between two points on the ocean floor. Diagrams A, B, and C show stages of sea-floor spreading. Spreading begins at A, continues to B where rocks at locations P begin to spread to the farther-apart positions we see in C. At C newer rock at the ocean crest S is dated at 10 million years.

Using the scale: 1mm = 50km, use a ruler on C to find:

1. The separation rate of the two continental landmasses in the past 10 million years, in cm/yr _____

2. The age of the sea floor at P in diagram C (1cm/yr = 10km/million years) _____

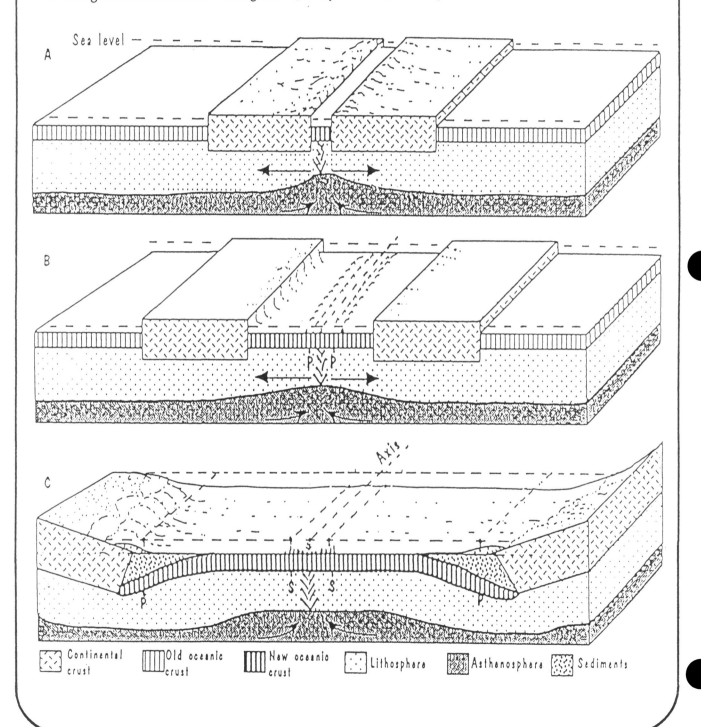

CONCEPTUAL PHYSICAL SCIENCE **EXPLORATIONS**
Chapter 34   Water on our World
Groundwater Flow and Contaminant Transport

The occupants of Houses 1, 2, and 3 wish to drill wells for domestic water supply. Note that the locations of all houses are between lakes A and B, at different elevations.

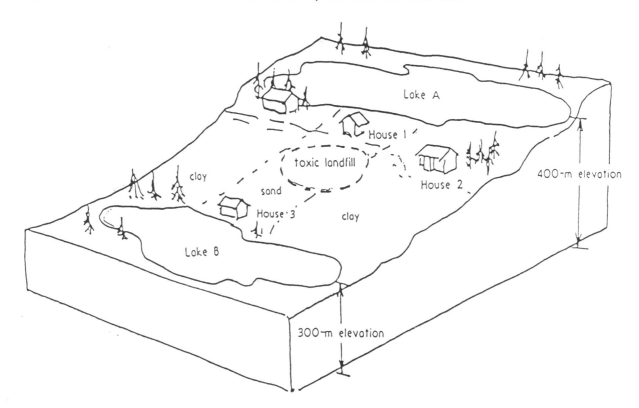

1. Show by sketching dashed lines on the drawing, the likely direction of groundwater flow beneath all the houses.

2. Which of the wells drilled beside Houses 1, 2, and 3 are likely to yield an abundant water supply?

_____

3. Do any of the three need to worry about the toxic landfill contaminating their water supply? Explain.

_____

4. Why don't the homeowners simply take water directly from the lakes?

_____

5. Suggest a potentially better location for the landfill. Defend your choice.

_____

_____

_____

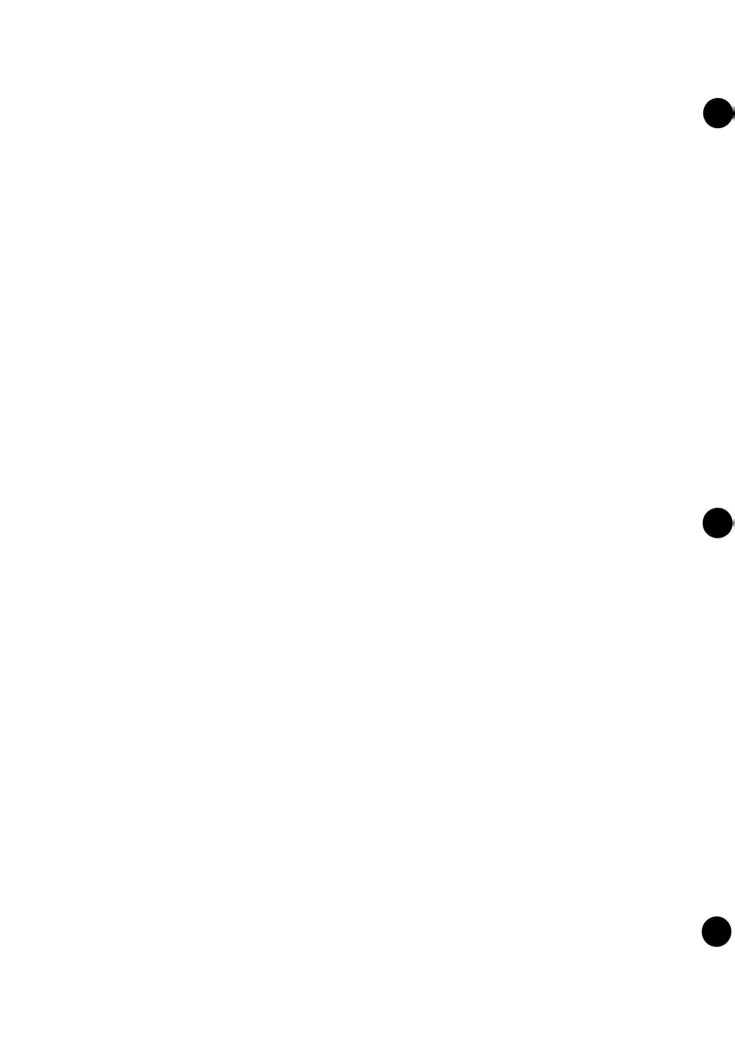

CONCEPTUAL PHYSICAL SCIENCE **EXPLORATIONS**

Chapter 35   Our Natural Landscape
Stream Flow

The diagram below illustrates a stream. Deposition of sediment occurs in one area, and erosion of sediment occurs in another area. On the diagram below mark the areas of deposition and the areas of erosion.

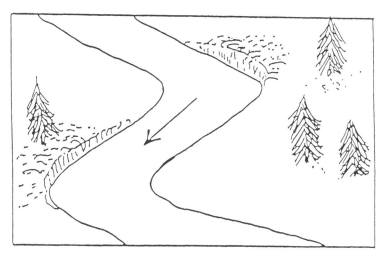

Farther downstream, the shape of the stream channel changes. Draw a likely new shape in the box below.

Place these words where they belong in the blanks below:

(*coarse-grained   increase   decreases   fine-grained*)

As the stream continues to meander, it widens the stream's valley into a broad, low-lying area called a *floodplain*. Floodplains are so named because they are the sites of periodic flooding. In a flood, as discharge and flow speed _____, so does the stream's ability to carry sediment. So, when a stream overflows its banks, sediment-rich water spills out onto the floodplain. Because the speed of the water quickly _____ as it spreads out over the large, flat, floodplain, a sequence of coarse to fine particles is deposited. Along the edges of the channel we can expect to find _____ sediment. Farther away from the stream channel, on the floodplain, we can expect to find _____ sediments.

Chapter 35 Our Natural Landscape
Stream Velocity

Let's explore how the average velocity of streams and rivers can change. Recall in Chapter 35 that the volume of water that flows past a given location over any given length of time depends both on the stream velocity and the cross-sectional area of the stream. We say

$$Q = A \times V$$

where Q is the volumetric flow rate, A is the cross sectional area of the stream, and V its average velocity.

Consider the stream shown below, with rectangular cross sectional areas

$A = \text{width} \times \text{depth}$

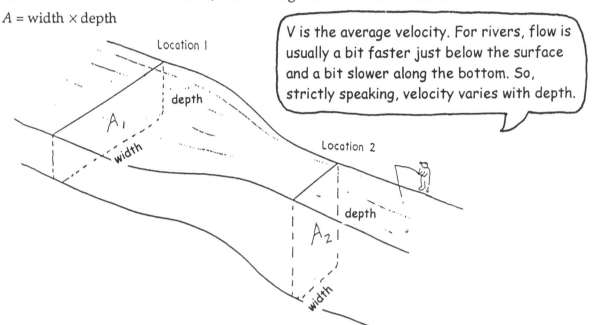

V is the average velocity. For rivers, flow is usually a bit faster just below the surface and a bit slower along the bottom. So, strictly speaking, velocity varies with depth.

1. The two locations shown have no inlets or outlets between them, so Q remains constant. Suppose the cross-sectional areas are also constant ($A_1 = A_2$), with Location 2 deeper but narrower than Location 1. What change, if any, occurs for the stream velocity?

_____

2. If Q remains constant, what happens to stream velocity at Location 2 if A_2 is less than A_1?

_____

3. If Q remains constant, what happens to stream velocity at Location 2 if A_2 is greater than A_1?

_____

4. What happens to stream velocity at Location 2 if area A_2 remains the same, but Q increases (perhaps by an inlet along the way?)

_____

5. What happens to stream velocity at Location 2 if both A_2 and Q increase?

_____

CONCEPTUAL PHYSICAL SCIENCE **EXPLORATIONS**

Chapter 35   Our Natural Landscape
Glacial Movement

From season to season the mass of a glacier changes. With each change in mass, the glacier moves. Glacier movement is measured by placing a line of markers across the ice and recording their changes in position over a period of time.

In the example below, we show the initial position of a line of markers and the glaciers terminus (the end of the glacier).

Draw the markers and glacier terminus at a later time for each of the following scenarios.

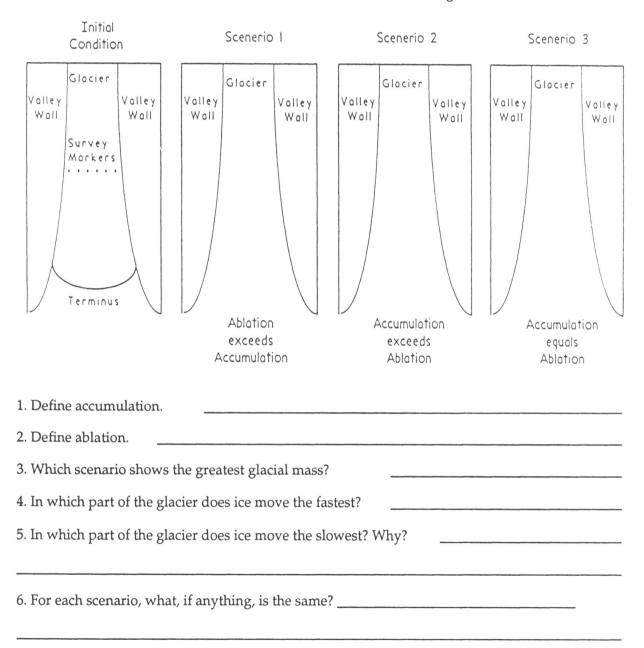

1. Define accumulation. _____

2. Define ablation. _____

3. Which scenario shows the greatest glacial mass? _____

4. In which part of the glacier does ice move the fastest? _____

5. In which part of the glacier does ice move the slowest? Why? _____

_____

6. For each scenario, what, if anything, is the same? _____

_____

CONCEPTUAL PHYSICAL SCIENCE **EXPLORATIONS**
Chapter 36 A Brief History of the Earth
Relative Time — What Came First?

The cross section below depicts many geologic events. In the space to the right, list the different events starting with the oldest to the youngest event. Where appropriate, include tectonic events (such as folding, deposition of beds, subsidence, uplift, erosion, intrusion).

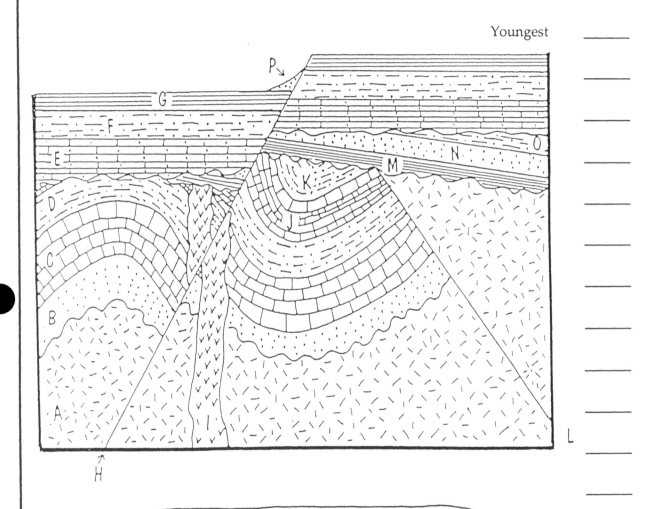

Youngest _____

_____

_____

_____

_____

_____

_____

_____

_____

_____

_____

_____

_____

_____

_____

_____

_____

_____

Oldest _____

Examine the rings in the cross section of a tree and you do more than determine the age of the tree. Relative thicknesses of the rings tells a lot about the climate conditions throughout the tree's history. A geologist similarly learns much about the Earth's history by examination of rock layers in cross sections of the Earth's crust.

From your investigation of the 6 geologic regions shown, answer the questions below.

The number of each question refers to the same-numbered region.

 Granite Basalt Plutonic rock Limestone Shale Sandstone

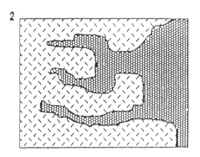

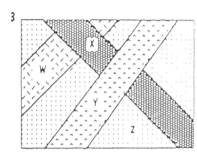

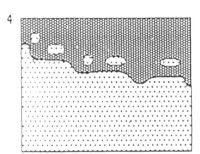

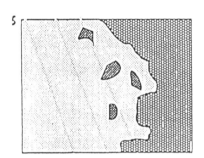

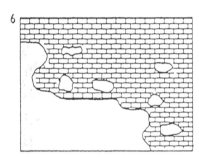

1. The shale has been cut by a dike. The radiometric age of the dike is estimated at 40 million years. Is the shale younger or older? _____

2. Which is older, the granite or the basalt? _____

3. The sandstone bed, Z, has been intruded by dikes. What is the age succession of dikes, going from oldest to youngest? _____

4. Which is older, the shale or the basalt? _____

5. Which is older, the sandstone or the basalt? _____

6. Which is older, the sandstone or the limestone? _____

CONCEPTUAL PHYSICAL SCIENCE **EXPLORATIONS**

Chapter 36   A Brief History of the Earth
Unconformities and Age Relationships

The wavy lines in the 4 regions below represent unconformities. Investigate the regions and answer corresponding questions about the 4 regions below.

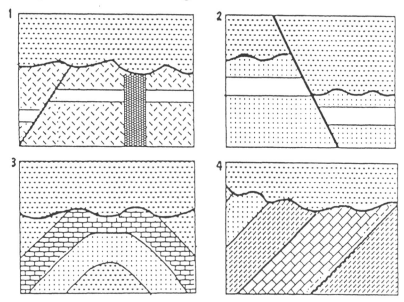

1. Did the faulting and dike occur before or after the unconformity? _____
   What kind of unconformity is it? _____

2. Did the faulting occur before or after the unconformity? _____

3. Did the folding occur before or after the unconformity? _____
   What kind of unconformity is it? _____

4. What kind of unconformity is represented? _____

5. Interestingly, the age of the Earth is some 4.5 billion years old—yet the oldest rocks found are
   some 3.8 billion years old. Why do we find no 4.5-billion year old rocks?

   _____

   _____

6. What is the age of the innermost ring in a living redwood tree that is 2000 years old? What is
   the age of the outermost ring? How does this example relate to the previous question?

   _____

   _____

7. What is the approximate age of the atoms that make up a 3.8-billion year old rock?

   _____

Chapter 36 A Brief History of the Earth
Our Earth's Hot Interior

A major puzzle faced scientist in the 19th century. Volcanoes showed that the Earth's interior is semi-molten. Penetration into the crust by bore-holes and mines showed that the Earth's temperature increases with depth. Scientists knew that heat flows from the interior to the surface. They assumed that the source of the Earth's internal heat was primordial, the afterglow of its fiery birth. Measurements of the Earth's rate of cooling indicated a relatively young Earth—some 25 to 30 million years in age. But geological evidence indicated an older Earth. This puzzle wasn't solved until the discovery of radioactivity. Then it was learned that the interior was kept hot by the energy of radioactive decay. We now know the age of the Earth is some 4.5 billion years—a much older Earth.

All rock contains trace amounts of radioactive minerals. Radioactive minerals in common granite release energy at the rate 0.03 J/kg·yr. Granite at the Earth's surface transfers this energy to the surroundings practically as fast as it is generated, so we don't find granite any warmer than other parts of our environment. But what if a sample of granite were thermally insulated? That is, suppose all the increase of thermal energy due to radioactive decay were contained. Then it would get hotter. How much hotter? Let's figure it out, using 790 J/kg·C° as the specific heat of granite.

Calculations to make:

1. How many joules are required to increase the temperature of 1 kg of granite by 500 C°?

_____

2. How many years would it take radioactivity in a kilogram of granite to produce this many joules?

_____

> Let's see now... back in Chapter 9 we learned that the relationship between quantity of heat, mass, specific heat and temperature difference is
> $$Q = cm\Delta T$$

Questions to answer:

1. How many years would it take a thermally insulated 1-kilogram chunk off granite to undergo a 500 C° increase in temperature?

_____

2. How many years would it take a thermally insulated one-million-kilogram chunk off granite to undergo a 500 C° increase in temperature?

_____

3. Why does the Earth's interior remain molten hot?

_____

4. Rock has a higher melting temperature deep in the interior. Why?

_____

> An electric toaster stays hot while electric energy is supplied, and doesn't cool until switched off. Similarly, do you think the energy source now keeping the Earth hot will one day suddenly switch off like a disconnected toaster - gradually decrease over a long time?

5. Why doesn't the Earth just keep getting hotter until it all melts?

_____

CONCEPTUAL PHYSICAL SCIENCE **EXPLORATIONS**
Chapter 36   A Brief History of the Earth
Radiometric Dating

Isotopes Commonly Used for Radiometric Dating		
Radioactive Parent	Stable Daughter Product	Half-life Value
Uranium-238	lead-206	4.5 billion years
Uranium-235	lead-207	704 million years
Potassium-40	argon-40	1.3 billion years
Carbon-14	nitrogen-14	5730 years

1. Consider a radiometric lab experiment wherein 99.98791 % of a certain radioactive sample of material remains after one year. What is the decay rate of the sample?

_____

2. What is the rate constant?
(Assume that the decay rate is constant for the one year period.)

_____

3. What is the half-life?

_____

4. Identify the isotope.

_____

5. In a sample collected in the field, this isotope was found to be 1/16 of its original amount. What is the age of the sample?

_____

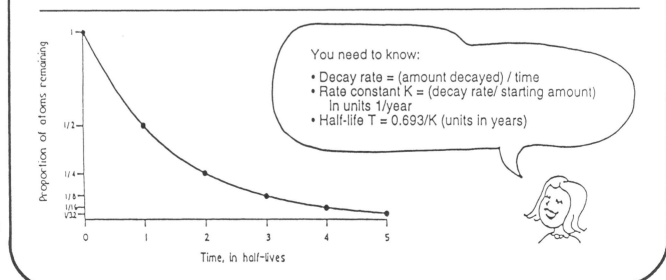

You need to know:

• Decay rate = (amount decayed) / time
• Rate constant K = (decay rate/ starting amount)
   In units 1/year
• Half-life T = 0.693/K (units in years)

CONCEPTUAL PHYSICAL SCIENCE EXPLORATIONS

Chapter 37   The Atmosphere, The Oceans, and Their Interactions
The Earth's Seasons

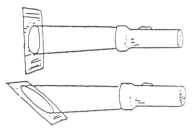

1. The warmth of equatorial regions and coldness of polar regions on the Earth can be understood by considering light from a flashlight striking a surface. If it strikes perpendicularly, light energy is more concentrated as it covers a smaller area; if it strikes at an angle, the energy spreads over a larger area. So the energy per unit area is less.

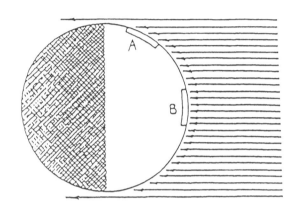

The arrows represent rays of light from the distant sun incident upon the Earth. Two areas of equal size are shown, Area A near the north pole and Area B near the equator. Count the rays that each reach each area, and explain why region B is warmer than region A.

2. The Earth's seasons result from the 23.5-degree tilt of the Earth's daily spin axis as it orbits the sun. When the Earth is at the position shown on the right in the sketch below (not shown to scale), the Northern Hemisphere tilts toward the sun, and sunlight striking it is strong (more rays per area). Sunlight striking the Southern Hemisphere is weak (fewer rays per area). Days in the north are warmer, and daylight lasts longer. You can see this by imagining the Earth making its complete daily 24-hour spin.

Do two things on the sketch: (1) Shade the Earth in nighttime darkness for all positions, as is already done in the left position. (2) Label each position with the proper month — March, June, September, or December.

Be sure to do the shading before you answer the questions on the backside of this sheet!

a. When the Earth is in any of the four positions shown, during one 24-hour spin a location at the equator receives sunlight half the time and is in darkness the other half of the time. This means that regions at the equator always get about _____ hours of sunlight and _____ hours of darkness.

b. Can you see that in the June position regions farther north have longer days and shorter nights? Locations north of the Arctic Circle (dotted in the Northern Hemisphere) are always illuminated by the sun as the Earth spins, so they get daylight _____ hours a day.

c. How many hours of light and darkness are there in June at regions south of the Antarctic Circle (dotted line in Southern Hemisphere)?

_____

d. Six months later, when the Earth is at the December position, is the situation in the Antarctic the same or is it the reverse?

_____

e. Why do South America and Australia enjoy warm weather in December instead of in June?

_____

3. The Earth spins about its polar axis once each 24 hours, which gives us day and night. If the Earth's spin was instead only one rotation per year, what difference would there be with day and night as we enjoy them now?

_____

_____

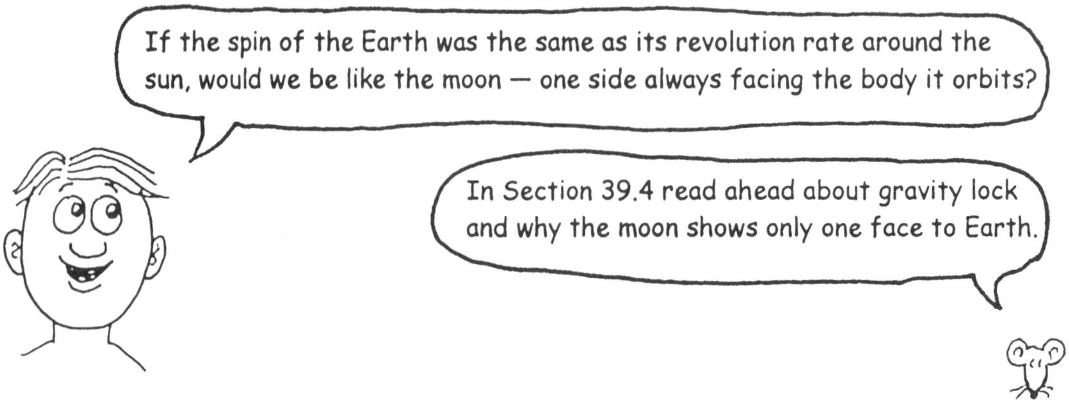

If the spin of the Earth was the same as its revolution rate around the sun, would we be like the moon — one side always facing the body it orbits?

In Section 39.4 read ahead about gravity lock and why the moon shows only one face to Earth.

130

CONCEPTUAL PHYSICAL SCIENCE **EXPLORATIONS**

Chapter 37   The Atmosphere, The Oceans, and Their Interactions
Short and Long Wavelengths

The sine curve is a pictorial representation of a wave—the high points being crests, and the low points troughs. The height of the wave is its *amplitude*. The wavelength is the distance between successive identical parts of the wave (like between crest to crest, or trough to trough). Wavelengths of water waves at the beach are measured in meters, wavelengths of ripples in a pond are measured in centimeters, and the wavelengths of light in billionths of a meter (nanometers).

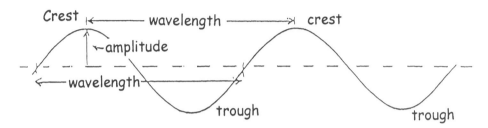

In the boxes below sketch three waves of the same amplitude—Wave A with half the wavelength of Wave B, and Wave C with wavelength twice as long as Wave B.

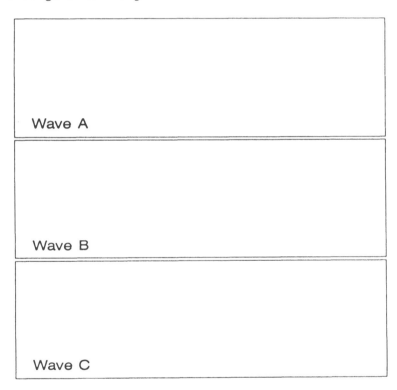

Wave A

Wave B

Wave C

1. If all three waves have the same speed, which has the highest frequency? _____

2. Compared with solar radiation, terrestrial radiation has a _____ wavelength.

3. In a florist's greenhouse, _____ waves are able to penetrate the greenhouse glass, but _____ waves cannot.

4. The Earth's atmosphere is similar to the glass in a greenhouse. If the atmosphere were to contain excess amounts of water vapor and carbon dioxide, the air would be opaque to _____ waves.

Chapter 37 The Atmosphere, The Oceans, and Their Interactions
Driving Forces of Air Motion

The primary driving force of the Earth's weather is _____. The unequal distribution of solar radiation on the Earth's surface creates temperature differences which in turn result in pressure differences in the atmosphere. These pressure differences generate horizontal winds as air moves from _____ pressure to _____ pressure. The weather patterns are not strictly horizontal though, there are other forces affecting the movement of air. Recall from Newton's second law that an object moves in the direction of the *net* force acting on it. The forces acting on the movement of air include:

    1) pressure gradient force 2) Coriolis force 3) centripetal force, and 4) friction

The greater the pressure difference the greater the force, and the greater the wind. The "push" caused by the horizontal differences in pressure across a surface is called the *pressure gradient force*. This force is represented by isobars on a weather map. Isobars connect locations on a map that have equal atmospheric pressure. The pressure gradient force is perpendicular to the isobars and strongest where the isobars are closely spaced. So,

the steeper the pressure gradient, the _____ the wind.

The *Coriolis force* is a result of the Earth's rotation. The Coriolis force is the deflection of the wind from a _____ path to a _____ path. The Coriolis force causes the wind to veer to the right of its path in the Northern Hemisphere and to the left of its path in the Southern Hemisphere.

As the wind blows around a low or high pressure center it constantly changes its direction. A change in speed or direction is acceleration. In order to keep the wind moving in a circular path the net force must be directed inward. This _____ force is called *centripetal force*.

The forces described above greatly influence the flow of upper winds (winds not influenced by surface frictional forces). The interaction of these forces cause the winds in the Northern Hemisphere to rotate _____ around regions of high pressure and _____ around regions of low pressure. In the Southern Hemisphere the situation is reversed — winds rotate _____ around a high and _____ around a low.

Winds blowing near the Earth's surface are slowed by *frictional forces*. In the Northern Hemisphere surface winds blow in a direction _____ into the centers of a low pressure area and _____ out of the centers of a high pressure area. The spiral direction is reversed in the Southern Hemisphere. Draw arrows to show the direction of the pressure gradient force.

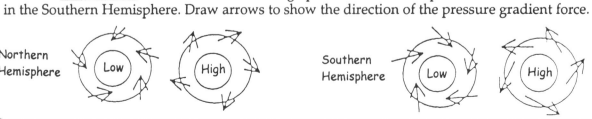

CONCEPTUAL PHYSICAL SCIENCE **EXPLORATIONS**

Chapter 38   Weather
Air Temperature and Pressure Patterns

Temperature patterns on weather maps are depicted by isotherms—lines that connect all points having the same temperature. Each isotherm separates temperatures having higher values from temperatures having lower values.

The following weather map to the right shows temperatures in degrees Fahrenheit for various locations. Using 10 degree intervals, connect same value numbers to construct isotherms. Label the temperature value at each end of the isotherm.
One isotherm has been completed as an example.

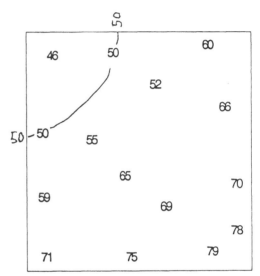

Tips for drawing Isotherms
- Isotherms can never be open ended.
- Isotherms are "closed" if they reach the boundary of plotted data, or make a loop.
- Isotherms can never touch, cross, or fork.
- Isotherms must always appear in sequence; for example, there must be a 60° isotherm between a 50- and 70-degree isotherm.
- Isotherms should be labeled with their values.

Pressure patterns on weather maps are depicted by isobars—lines which connect all points having equal pressure. Each isobar separates stations of higher pressure from stations of lower pressure.

The weather map below shows air pressure in millibar (mb) units at various locations. Using an interval of 4 (for example, 1008, 1012, 1016 etc.), connect equal pressure values to construct isobars. Label the pressure value at each end of the isobar. One isobar has been completed as an example.

- Tips for drawing isobars are similar to those for drawing isotherms.

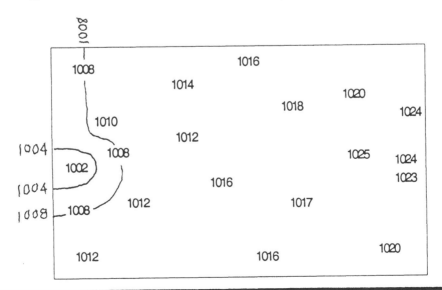

Air Temperature and Pressure Patterns continued:

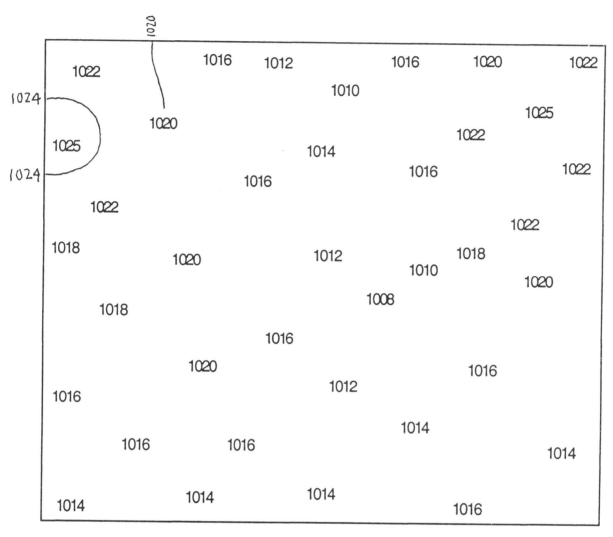

On the map above, use an interval of 4 to draw lines of equal pressure (isobars) to show the pattern of air pressure. Locate and mark regions of high pressure with an "H" and regions of low pressure with an "L". Use the map to answer the questions below.

1. On the map above, areas of high pressure are depicted by the _____ isobar.

2. On the map above, areas of low pressure are depicted by the _____ isobar.

Circle the correct answer

3. Highs are usually accompanied by (stormy weather) (fair weather).

4. In the Northern Hemisphere, surface winds surrounding a high pressure system blow in a (clockwise direction) (counterclockwise direction).

5. In the Northern Hemisphere, surface winds spiral inward into a (region of low pressure) (region of high pressure).

CONCEPTUAL PHYSICAL SCIENCE **EXPLORATIONS**
Chapter 38: Weather
Surface Weather Maps

Station models are used on weather maps to depict weather conditions for individual localities. Weather codes are plotted in, on, and around a central circle that describes the overall appearance of the sky. Jutting from the circle is a wind arrow, its tail in the direction from which the windcomes and its feathers indicating the wind speed. Other weather codes are in standard position around the circle.

Use the simplified station model and weather symbols to complete the statements below.

Total Sky Cover

◯ No clouds

◔ Less than one-tenth or one-tenth

◑ Two-tenths or three-tenths

◕ Four-tenths

◐ Five-tenths

◒ Six-tenths

◕ Seven-tenths or eight-tenths

◗ Nine-tenths or overcast with openings

● Completely overcast

⊗ Sky obscured

Pressure Tendency

╱ Rising, then falling

╱ Rising, then steady; or rising, then rising more slowly

╱ Rising steadily, or unsteadily

✓ Falling or steady, then rising; or rising, then rising more quickly

— Steady, same as 3 hours ago

∨ Falling, then rising, same or lower than 3 hours ago

╲ Falling, then steady; or falling, then falling more slowly

╲ Falling steadily, or unsteadily

╱╲ Steady or rising, then falling; or falling, then falling more quickly

Barometer no higher than 3 hours ago

Barometer no lower than 3 hours ago

Wind Entries

	Miles (Statute) Per Hour	Knots	Kilometers Per Hour
◎	Calm	Calm	Calm
——	1-2	1-2	1-3
⌐	3-8	3-7	4-13
⌐	9-14	8-12	14-19
⌐	15-20	13-17	20-32
⌐	21-25	18-22	33-40
⌐	26-31	23-27	41-50
⌐	32-37	28-32	51-60
⌐	38-43	33-37	61-69
⌐	44-49	38-42	70-79
⌐	50-54	43-47	80-87
⌐	55-60	48-52	88-96
⌐	61-66	53-57	97-106
⌐	67-71	58-62	107-114
⌐	72-77	63-67	115-124
⌐	78-83	68-72	125-143

Common Weather Symbols

• Light rain   ▽ Rain shower

•• Moderate rain   ▽ Snow shower

•• Heavy rain   △ Showers of hail

✱ ✱ Light snow   ↟ Drifting or blowing snow

✱✱ Moderate snow   ↶ Dust storm

✱✱ Heavy snow   ≡ Fog

❞❞ Light drizzle   ∞ Haze

△ Ice pellets (sleet)   ∿ Smoke

⊙ Freezing rain   ⎰ Thunderstorm

⊙ Freezing drizzle   ⚡ Hurricane

Wind speed

Wind direction

Barometric pressure reduced to sea level

Temperature (F°)
31

Present Weather
★

Visiblility
24

Dew Point
30

250

Pressure higher or lower than 3 hours ago   30

+28   Barometric tendency in last 3 hours

Amount of change during last 3 hours
4

Time precipitation began or ended

Weather during past 6 hours

45

Amount of precipitation during past 6 hours

1. The overall appearance of the sky is _____
.

2. The wind speed is _____ kilometers per hour.

3. The wind direction is coming from the _____

4. The present weather conditions call for _____

5. The barometric tendency is _____

6. For the past 6 hours the weather conditions have been _____

Surface Weather Maps continued:

Use the unlabeled station model to answer the questions below.

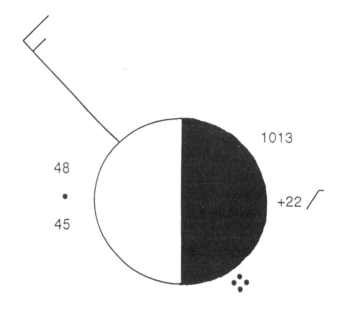

1. The overall appearance of the sky is _____ .

2. The wind speed is _____ kilometers per hour.

3. The wind direction is coming from the _____ .

4. The present weather conditions call for _____ .

5. The barometric tendency is _____ .

6. For the past 6 hours the weather conditions have been _____ .

7. The barometric pressure is _____ .

8. The dew point is _____ .

9. The current temperature is _____ .

10. Compared to the past few hours, barometric pressure is _____ .

CONCEPTUAL PHYSICAL SCIENCE EXPLORATIONS

Chapter 39 The Solar System
Earth-Moon-Sun Alignments

You cast a shadow whenever you stand in the
sunlight. Everything does, including planetary
bodies. To better understand this, consider the
sketch of the shadow cast by the apple. Note how the
rays define the darkest part of the shadow, the
umbra, and the lighter part, the *penumbra*. During a solar eclipse, the moon similarly casts a
shadow on the Earth. The region of "totality" is the umbra, and regions getting a partial
eclipse are in the penumbra.

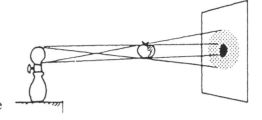

    Below is a diagram of the sun, Earth, and the orbital path of the moon (dashed circle). One
position of the moon is shown. Draw the moon in the appropriate positions on the dashed
circle to represent (a) a quarter moon; (b) a half moon; (c) a solar eclipse; (d) a lunar eclipse.
Label your positions. For c and d, extend rays from the top and bottom of the sun to show
umbra and penumbra regions on the Earth.

Eclipses are relatively rare because the orbital planes of the
Earth about the sun and the moon about the Earth are
slightly tipped to each other — so the shadows usually
"miss." If the planes weren't tipped, eclipses would occur
monthly. Can you see that eclipses only occur when the
three bodies align along the intersection of these planes
(points A and B)?

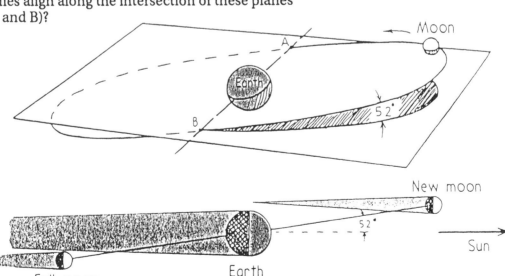

Because of the large size of the sun, its rays taper as shown. Note that the shadow of the new
moon misses the Earth, and the shadow of the Earth misses the full moon. Would these
shadows miss when alignment of the three bodies is along points A and B? _____

Earth-Moon-Sun Alignment continued:

Sketch the appropriate positions of the moon in its orbit about the Earth for (a) a solar eclipse; (b) a lunar eclipse. Label your positions. Sketch solar rays similar to those you constructed on the previous page.

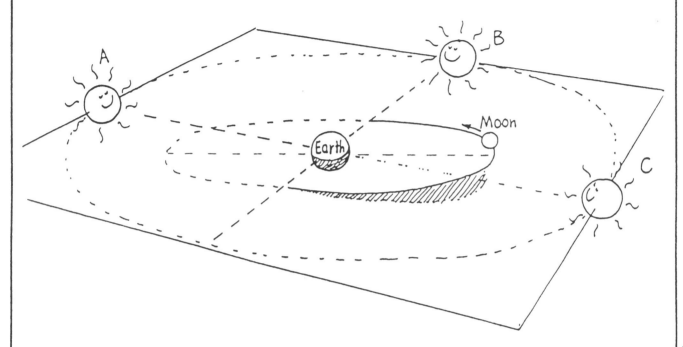

Shown below are (a) a partial solar eclipse in progress, and (b) a partial lunar eclipse in progress. Fill in the blanks and label the correct phases of the moon at these times (New moon, quarter moon, full moon, etc.).

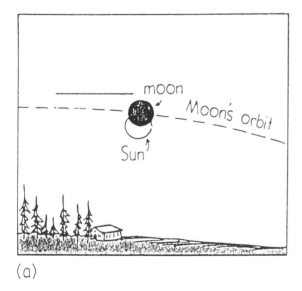

(a)

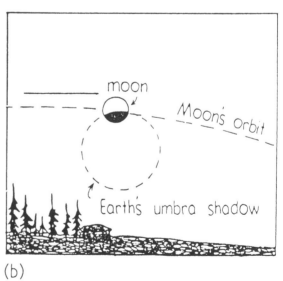

(b)

CONCEPTUAL PHYSICAL SCIENCE EXPLORATIONS

Chapter 40 The Stars
Stellar Parallax

Finding distances to objects beyond the solar system is based on the simple phenomenon of **parallax**. Hold a pencil at arm's length and view it against a distant background — each eye sees a different view (try it and see). The displaced view indicates distance. Likewise, when the Earth travels around the sun each year, the position of relatively nearby stars shifts slightly relative to the background stars. By carefully measuring this shift, astronomer types can determine the distance to nearby stars.

Can you see why the close star appears to shift positions relative to the background stars? And how maximum shift appears in observations 6 months apart?

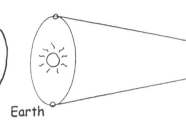

The photographs below show the same section of the evening sky taken at a 6-month interval. Investigate the photos carefully and determine which star appears in a different positioin relative to the others. Circle the star that shows a parallax shift.

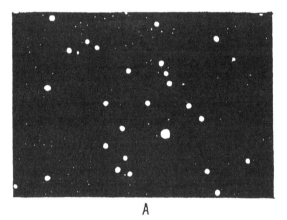
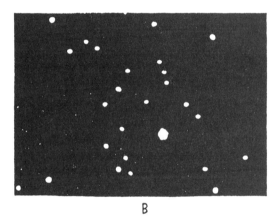

A   B

Below are three sets of photographs, all taken at 6-month intervals. Circle the stars that show a parallax shift in each of the photos.

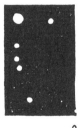

Set A   Set B   Set C

Use a fine ruler and measure the distance of shift in millimeters and place the values below:

Set A _____ mm   Set B _____   Set C _____ mm

Which set of photos indicate the closest star? The most distant "parallaxed" star?.

CONCEPTUAL PHYSICAL SCIENCE EXPLORATIONS
Appendix C Vectors

Vectors and the Parallelogram Rule

1. When vectors A and B are at an angle to each other, they add to produce the resultant C by the *parallelogram rule*. Note that C is the diagonal of a parallelogram where A and B are adjacent sides. Resultant C is shown in the first two diagrams, *a* and *b*. Construct the resultant C in diagrams *c* and *d*. Note that in diagram *d* you form a rectangle (a special case of a parallelogram).

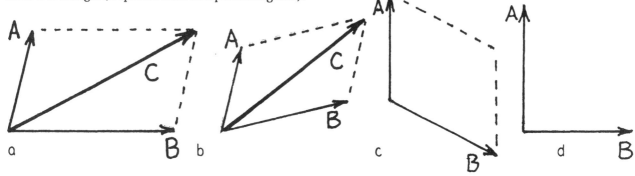

2. Below we see a top view of an airplane being blown offcourse by wind in various directions. Use the parallelogram rule to show the resulting speed and direction of travel for each case. In which case does the airplane travel fastest across the ground? _____ Slowest? _____

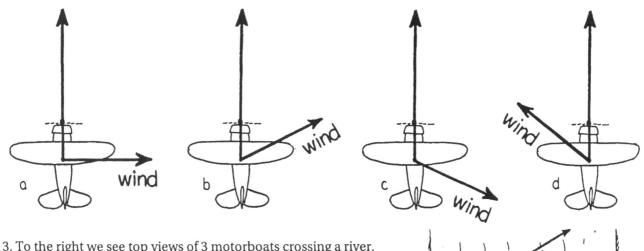

3. To the right we see top views of 3 motorboats crossing a river. All have the same speed relative to the water, and all experience the same water flow.

   Construct resultant vectors showing the speed and direction of the boats.

   a. Which boat takes the shortest path to the opposite shore? _____

   b. Which boat reaches the opposite shore first? _____

   c. Which boat provides the fastest ride? _____

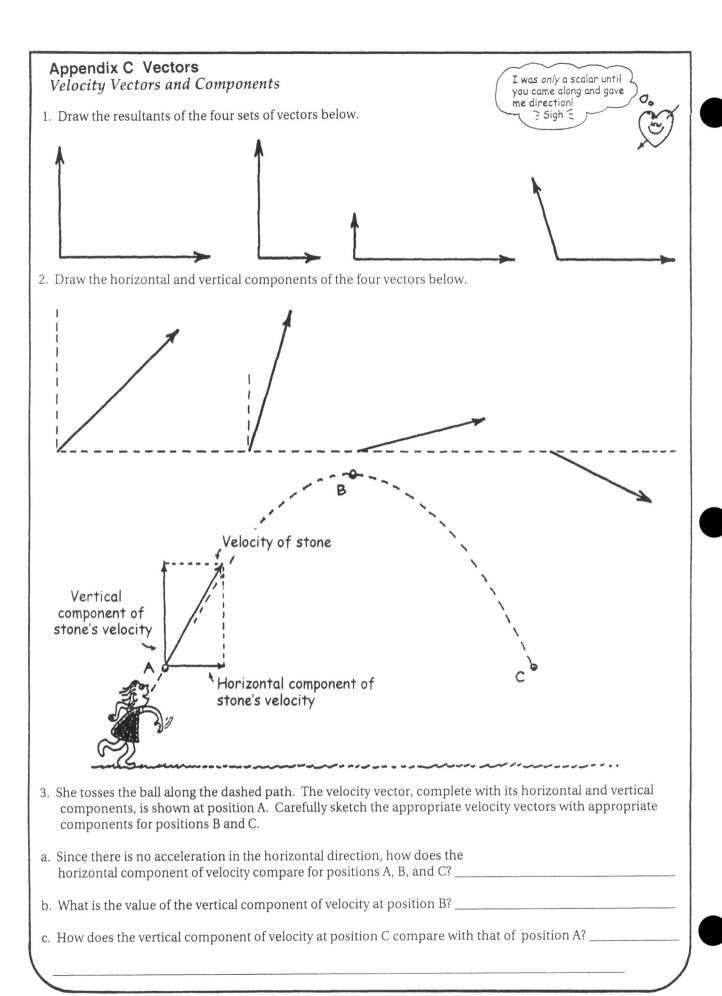

Appendix C Vectors
Velocity Vectors and Components

I was *only* a scalar until you came along and gave me direction!
⋛ Sigh ⋚

1. Draw the resultants of the four sets of vectors below.

2. Draw the horizontal and vertical components of the four vectors below.

Velocity of stone

Vertical component of stone's velocity

A

Horizontal component of stone's velocity

B

C

3. She tosses the ball along the dashed path. The velocity vector, complete with its horizontal and vertical components, is shown at position A. Carefully sketch the appropriate velocity vectors with appropriate components for positions B and C.

a. Since there is no acceleration in the horizontal direction, how does the horizontal component of velocity compare for positions A, B, and C? _____

b. What is the value of the vertical component of velocity at position B? _____

c. How does the vertical component of velocity at position C compare with that of position A? _____

_____

CONCEPTUAL PHYSICAL SCIENCE EXPLORATIONS
Appendix C Vectors
Vectors and Sailboats

(Do not attempt this until you have studied Appendix C!)

1. The sketch shows a top view of a small railroad car pulled by a rope. The force *F* that the rope exerts on the car has one component along the track, and another component perpendicular to the track.

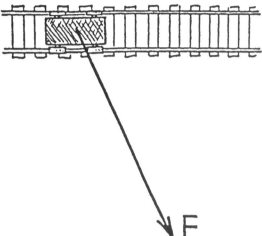

   a. Draw these components on the sketch. Which component is larger?

   _____

   b. Which component produces acceleration?

   _____

   c. What would be the effect of pulling on the rope if it were perpendicular to the track?

   _____

2. The sketches below represent simplified top views of sailboats in a cross-wind direction. The impact of the wind produces a FORCE vector on each as shown.
    (We do NOT consider *velocity* vectors here!)

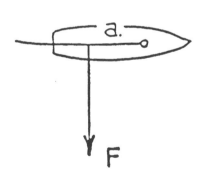

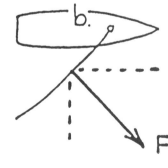

   a. Why is the position of the sail above useless for propelling the boat along its forward direction? (Relate this to Question 1c. above. Where the train is constrained by tracks to move in one direction, the boat is similarly constrained to move along one direction by its deep vertical fin — the *keel*.)

   _____

   _____

   b. Sketch the component of force parallel to the direction of the boat's motion (along its keel), and the component perpendicular to its motion. Will the boat move in a forward direction? (Relate this to Question 1b. above.)

   _____

   _____

3. The boat to the right is oriented at an angle into the wind. Draw the force vector and its forward and perpendicular components.

a. Will the boat move in a forward direction and tack into the wind? Why or why not?

_____

_____

4. The sketch below is a top view of five identical sailboats. Where they exist, draw force vectors to represent wind impact on the sails. Then draw components parallel and perpendicular to the keels of each boat.

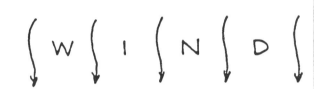

a. Which boat will sail the fastest in a forward direction?

_____

b. Which will respond least to the wind?

_____

c. Which will move in a backward direction?

_____

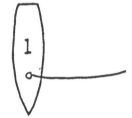

d. Which will experience less and less wind impact with increasing speed?

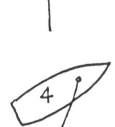

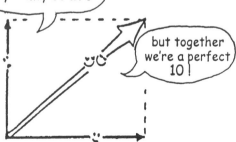

144

CONCEPTUAL PHYSICAL SCIENCE EXPLORATIONS

Appendix C Vectors

Force-Vector Diagrams

Being able to identify forces that act on a body is enormously important to any further study of physics. For example, in sketch 1 below we see two forces acting on the suspended rock — the force of gravity downward (W) and tension of the string (T) upward. These are the only forces that act on the rock. We see the rock in sketch 2 being pulled by two strings, so there are three forces, A, B, and C. The size of vectors A and B, relative to W, can be determined by the parallelogram rule. See if you can apply it to sketches 3 and 4. Continue identifying and sketching proper force vectors for the others. Good Energy!

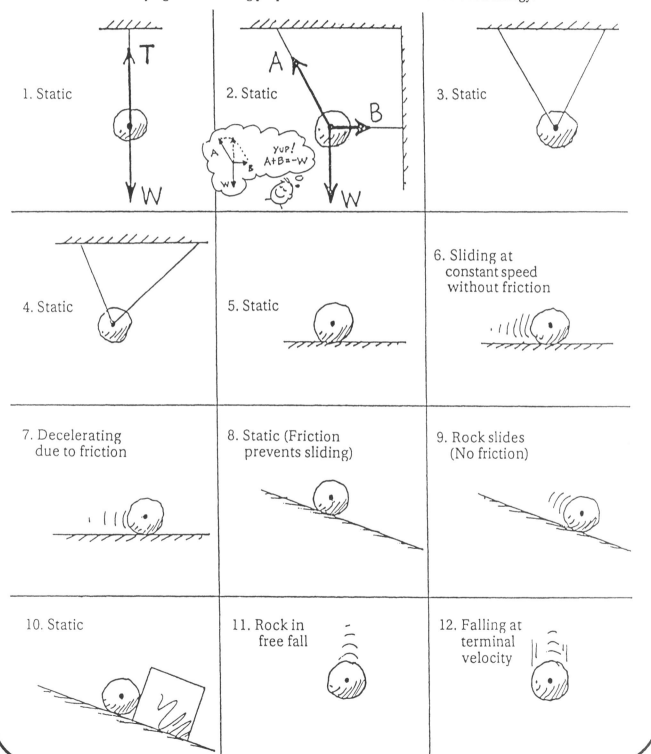

1. Static

2. Static

3. Static

4. Static

5. Static

6. Sliding at constant speed without friction

7. Decelerating due to friction

8. Static (Friction prevents sliding)

9. Rock slides (No friction)

10. Static

11. Rock in free fall

12. Falling at terminal velocity

CONCEPTUAL PHYSICAL SCIENCE EXPLORATIONS

Appendix D Fluid Physics

Archimedes' Principle I

1. Consider a balloon filled with 1 liter of water (1000 cm³) in equilibrium in a container of water, as shown in Figure 1.

   a. What is the mass of the 1 liter of water?

   _____

   b. What is the weight of the 1 liter of water?

   _____

   c. What is the weight of water displaced by the balloon?

   _____

   d. What is the buoyant force on the balloon?

   _____

   e. Sketch a pair of vectors in Figure 1: one for the weight of the balloon and the other for the buoyant force that acts on it. How do the size and directions of your vectors compare?

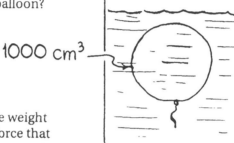

1000 cm³

Figure 1

2. As a thought experiment, pretend we could remove the water from the balloon but still have it remain the same size of 1 liter. Then inside the balloon is a vacuum.

   a. What is the mass of the liter of nothing?

   _____

   b. What is the weight of the liter of nothing?

   _____

   c. What is the weight of water displaced by the massless balloon?

   _____

   d. What is the buoyant force on the massless balloon?

   _____

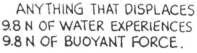

ANYTHING THAT DISPLACES 9.8 N OF WATER EXPERIENCES 9.8 N OF BUOYANT FORCE.

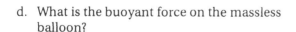

CUZ IF YOU PUSH 9.8 N OF WATER ASIDE THE WATER PUSHES BACK ON YOU WITH 9.8 N !

   e. In which direction would the massless balloon be accelerated?

   _____

Archimedes' Principle I continued:

3. Assume the balloon is replaced by a 0.5-kilogram piece of wood that has exactly the same volume (1000 cm³), as shown in Figure 2. The wood is held in the same submerged position beneath the surface of the water.

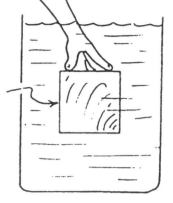

1000 cm³

   a. What volume of water is displaced by the wood?

   _____

   b. What is the mass of the water displaced by the wood?

   _____

   c. What is the weight of the water displaced by the wood?

   _____

Figure 2

   d. How much buoyant force does the surrounding water exert on the wood?

   _____

   e. When the hand is removed, what is the net force on the wood?

   _____

   f. In which direction does the wood accelerate when released? _____

THE BUOYANT FORCE ON A SUBMERGED OBJECT EQUALS THE WEIGHT OF WATER DISPLACED

... NOT THE WEIGHT OF THE OBJECT ITSELF!

... UNLESS IT IS FLOATING!

4. Repeat parts *a* through *f* in the previous question for a 5-kg rock that has the same volume (1000 cm³), as shown in Figure 3. Assume the rock is suspended by a string in the container of water.

   a. _____

   b. _____

   c. _____

   d. _____

   e. _____

   f. _____

Figure 3

1000 cm³

WHEN THE WEIGHT OF AN OBJECT IS GREATER THAN THE BUOYANT FORCE EXERTED ON IT, IT SINKS!

CONCEPTUAL PHYSICAL SCIENCE EXPLORATIONS

Appendix D Fluid Physics

Archimedes' Principle II

1. The water lines for the first three cases are shown. Sketch in the appropriate water lines for cases *d* and *e*, and make up your own for case *f*.

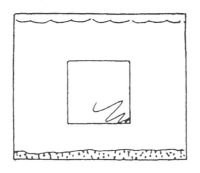
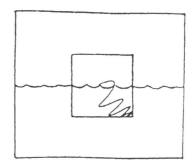

a. DENSER THAN WATER   b. SAME DENSITY AS WATER   c. 1/2 AS DENSE AS WATER

d. 1/4 AS DENSE AS WATER   e. 3/4 AS DENSE AS WATER   f._____AS DENSE AS WATER

2. If the weight of a ship is 100 million N, then the water it displaces weighs_____ .

    If cargo weighing 1000 N is put on board then the ship will sink down until an extra

    _____ of water is displaced.

3. The first two sketches below show the water line for an empty and a loaded ship. Draw in the appropriate water line for the third sketch.

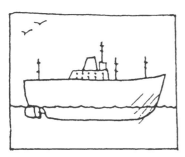

a. SHIP EMPTY   b. SHIP LOADED WITH 50   c. SHIP LOADED WITH 50
                            TONS OF IRON   TONS OF STYROFOAM

Archimedes' Principle II continued:

4. Here is a glass of ice water with an ice cube floating in it. Draw the water line after the ice cube melts. (Will the water line rise, fall, or remain the same?)

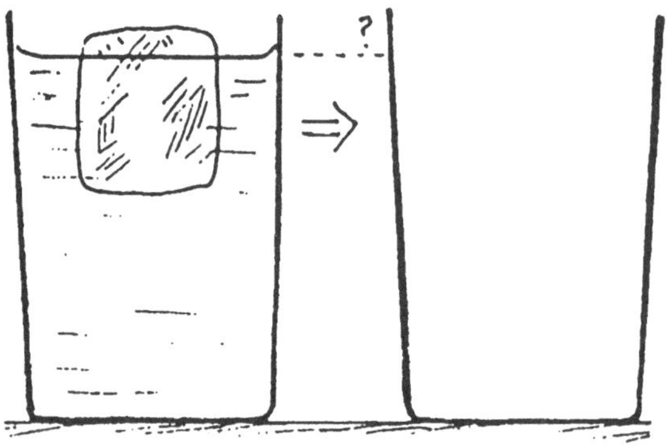

5. The air-filled balloon is weighted so it sinks in water. Near the surface, the balloon has a certain volume. Draw the balloon at the bottom (inside the dashed square) and show whether it is bigger, smaller, or the same size.

    a. Since the weighted balloon sinks, how does its overall density compare to the density of water?

    b. As the weighted balloon sinks, does its density increase, decrease, or remain the same?

    c. Since the weighted balloon sinks, how does the buoyant force on it compare to its weight?

    d. As the weighted balloon sinks deeper, does the buoyant force on it increase, decrease, or remain the same?

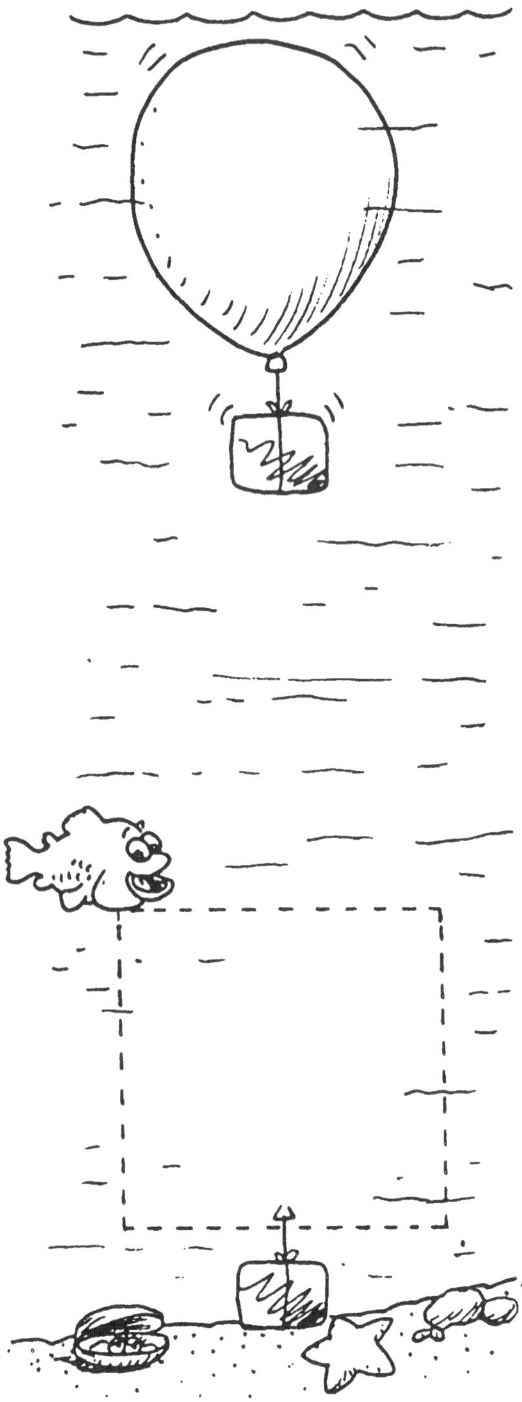

6. What would be your answers to Questions ***a***, ***b***, ***c***, and ***d*** for a rock instead of an air-filled balloon?

    a.

    b.

    c.

    d.

Appendix D Fluid Physics
Gases

1. A principle difference between a liquid and a gas is that when a liquid is under pressure, its volume

    (increases)   (decreases)   (doesn't change noticeably)

    and its density

    (increases)   (decreases)   (doesn't change noticeably)

    When a gas is under pressure, its volume

    (increases)   (decreases)   (doesn't change noticeably)

    and its density

    (increases)   (decreases)   (doesn't change noticeably)

2. The sketch shows the launching of a weather balloon at sea level. Make a sketch of the same weather balloon when it is high in the atmosphere. In words, what is different about its size and why?

    _____

    _____

    _____

    _____

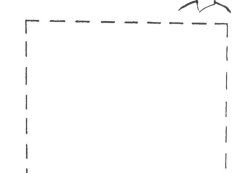

HIGH-ALTITUDE SIZE

GROUND-LEVEL SIZE

3. A hydrogen-filled balloon that weighs 10 N must displace_____N of air in order to float in air.

    If it displaces less than_____N it will be buoyed up with less than_____N and sink.

    If it displaces more than_____N of air it will move upward.

4. Why is the cartoon more humorous to physics types than to non-physics types? What physics has occurred?

    _____

    _____

    _____

    _____

RATS TO YOU TOO, DANIEL BERNOULLI !

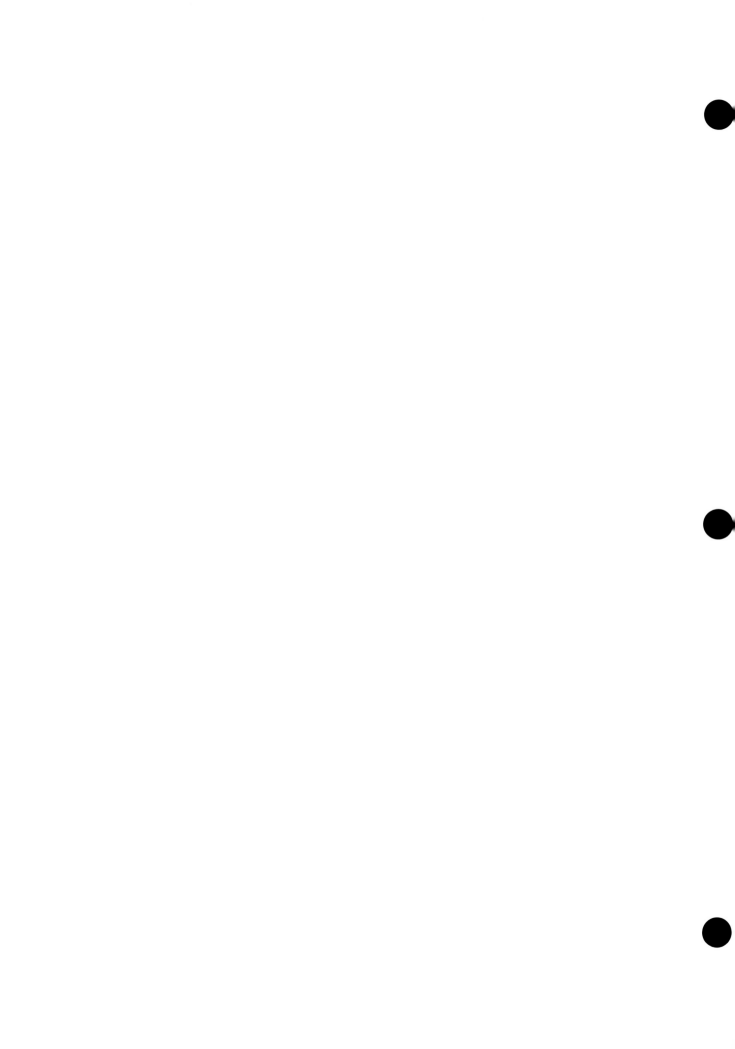

Answers to the Explorations Practice Pages

Compare your responses to the previous pages with our responses in
the reduced pages that follow. You have the choice of taking a shortcut
and looking at our responses first—or you can be nice to yourself and
work out your own without peeking ahead. In working through these
pages on your own or with friends, without looking ahead until you've
given each a good try, you may get that nice feeling that comes with
doing a good thing well.

CONCEPTUAL PHYSICAL SCIENCE EXPLORATIONS

Chapter 1 About Science
Making Hypotheses

The word science comes from Latin, meaning "to know."
The word *hypothesis* comes from Greek, "under an idea."
A hypothesis (an educated guess) often leads to new knowledge and may help to establish a theory.

WHICH IS AN EDUCATED GUESS... A HYPOTHESIS OR A THEORY?

...WHICH RESULTS FROM A LARGE BODY OF KNOWLEDGE?

Examples:

1. It is well known that most objects expand when heated. An iron plate gets slightly bigger, for example, when put in a hot oven. But what of a hole in the middle of the plate? Will the hole get bigger or smaller when expansion occurs? One friend may say the size of the hole will increase, and another says it will decrease.

I CUT A DISK FROM THIS IRON PLATE. WHEN I HEAT THE PLATE, WILL THE HOLE GET BIGGER, OR SMALLER?

a. What is your hypothesis about hole size? Is there a test for finding out?

HYP 1: HOLE GETS BIGGER. HYP 2: SMALLER. HYP 3: NO CHANGE.

TEST: HEAT IT IN AN OVEN, THEN MEASURE! (HYP 1 IS CORRECT.)

WHAT HAPPENS IF HE PLUGS THE HOLE INTO THE HOLE BEFORE HEATING EVERYTHING?

b. There are often several ways to test a hypothesis. For example, you can perform a physical experiment and witness the results yourself, or you can use the library to find the reported results of other investigators. Which of these two methods do you favor, and why?

(IT DEPENDS ON THE SITUATION—MOST RESEARCH INVOLVES BOTH.)

2. Before the time of the printing press, books were hand-copied by scribes, many of whom were monks in monasteries. There is the story of the scribe who was frustrated to find a smudge on an important page he was copying. The smudge blotted out part of the sentence that reported the number of teeth in the head of a donkey. The scribe was very upset and didn't know what to do. He consulted with other scribes to see if any of their books stated the number of teeth in the head of a donkey. After many hours of fruitless searching through the library, it was agreed that the best thing to do was to send a messenger by donkey to the next monastery and continue the search there. What would be your advice?

ACTUALLY LOOK IN THE DONKEY'S MOUTH AND COUNT! (WATCH FOR MISSING TEETH)

Making Distinctions

MATERIAL ARTIFACTS

BAN AUTOMOBILES

DOWN WITH TECHNOLOGY

Many people don't seem to see the difference between a thing and the *abuse* of the thing. For example, a city council that bans skateboarding may not distinguish between skateboarding and reckless skateboarding. A person who advocates that a particular technology be banned may not distinguish between that technology and the abuses of that technology. There's a difference between a thing and the abuse of the thing.

On a separate sheet of paper, list other examples where use and abuse are often not distinguished. Compare your list with others in your class.

CONCEPTUAL PHYSICAL SCIENCE EXPLORATIONS

Chapter 2 Newton's First Law—the Law of Inertia
Inertia

(Circle the correct answers.)

1. An astronaut in outer space away from gravitational or frictional forces throws a rock. The rock will

    (continue moving in a straight line at constant speed)

    (gradually slow to a stop)

    The rock's tendency to do this is called

    (inertia) (weight) (acceleration)

2.

    The sketch shows a top view of a rock being whirled at the end of a string (clockwise). If the string breaks, the path of the rock is

    (A) (B) (C) (D)

3. Suppose you are standing in the aisle of a bus that travels along a straight road at 100 km/h, and you hold a pencil still above your head. Then relative to the bus, the velocity of the pencil is 0 km/h, and relative to the road, the pencil has a horizontal velocity of

    (less than 100 km/h) (100 km/h) (more than 100 km/h)

    Suppose you release the pencil. While it is dropping, and relative to the road, the pencil still has a horizontal velocity of

    (less than 100 km/h) (100 km/h) (more than 100 km/h)

    This means that the pencil will strike the floor at a place directly

    (behind you) (at your feet below your hand) (in front of you)

    Relative to you, the way the pencil drops

    (is the same as if the bus were at rest)

    (depends on the velocity of the bus)

    How does this example illustrate the law of inertia?
    A body in motion tends to remain in motion as long as no net force is exerted on the body in the direction of motion. Since there is no horizontal force on the pencil, its horizontal motion doesn't change.

CONCEPTUAL PHYSICAL SCIENCE EXPLORATIONS

Chapter 2 Newton's First Law—the Law of Inertia
The Equilibrium Rule: $\Sigma F = 0$

1. Little Nellie Newton wishes to be a gymnast and hangs from a variety of positions as shown. Since she is not accelerating, the net force on her is zero. This means the upward pull of the rope(s) equals the downward pull of gravity. She weighs 300 N. Show the scale reading for each case.

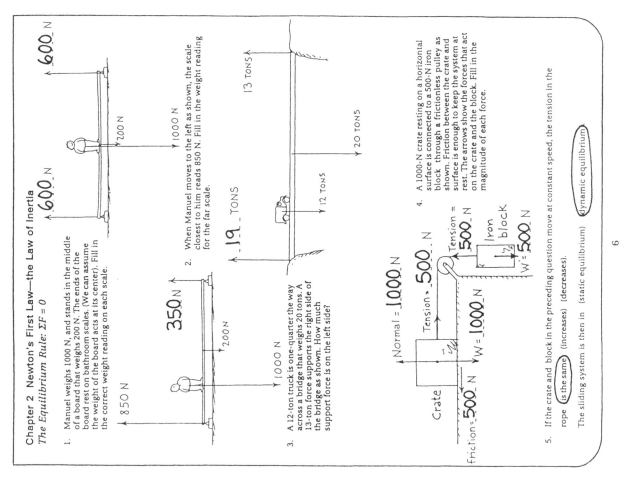

300 N · 300 N / 300 N · 150 N

100 N · 150 N · 300 N

2. When Burl the painter stands in the exact middle of his staging, the left scale reads 600 N. Fill in the reading on the right scale. The total weight of Burl and staging must be **1200** N.

600 N · 600 N

3. Burl stands farther from the left. Fill in the reading on the right scale.

400 N · 800 N

4. In a silly mood, Burl dangles from the right end. Fill in the reading on the right scale.

0 N · 1200 N

Chapter 2 Newton's First Law—the Law of Inertia
The Equilibrium Rule: $\Sigma F = 0$

1. Manuel weighs 1000 N, and stands in the middle of a board that weighs 200 N. The ends of the board rest on bathroom scales. (We can assume the weight of the board acts at its center). Fill in the correct weight reading on each scale.

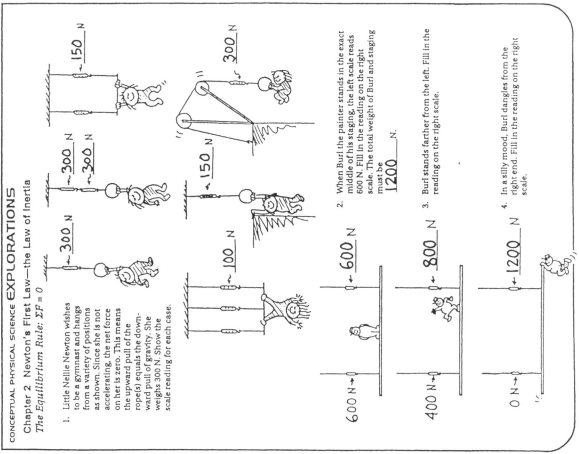

600 N · 600 N

2. When Manuel moves to the left as shown, the scale closest to him reads 850 N. Fill in the weight reading for the far scale.

850 N · 350 N · 200 N · 1000 N

3. A 12-ton truck is one-quarter the way across a bridge that weighs 20 tons. A 13-ton force supports the right side of the bridge as shown. How much support force is on the left side?

19 TONS

1000 N · 13 TONS · 12 TONS · 20 TONS

4. A 1000-N crate resting on a horizontal surface is connected to a 500-N iron block through a frictionless pulley as shown. Friction between the crate and surface is enough to keep the system at rest. The arrows show the forces that act on the crate and the block. Fill in the magnitude of each force.

Normal = **1000** N

Tension = **500** N

friction = **500** N · W = **1000** N

Crate

Tension = **500** N

Iron block · W = **500** N

5. If the crate and block in the preceding question move at constant speed, the tension in the rope (is the same) (increases) (decreases).

The sliding system is then in (static equilibrium) (dynamic equilibrium).

CONCEPTUAL PHYSICAL SCIENCE EXPLORATIONS
Chapter 2, 3, or Appendix B: Motion
Free Fall Speed

1. Aunt Minnie gives you $10 per second for 4 seconds. How much money do you have after 4 seconds? **$40**

2. A ball dropped from rest picks up speed at 10 m/s per second. After 4 seconds, how fast is it falling? **40 m/s**

3. You have $20, and Uncle Harry gives you $10 each second for 3 seconds. How much money do you have after 3 seconds? **$50**

4. A ball is thrown straight down with an initial speed of 20 m/s. After 3 seconds, how fast is it falling? **50 m/s**

5. You have $50 and you pay Aunt Minnie $10/second. When will your money run out? **5 s**

6. You shoot an arrow straight up at 50 m/s. When will it run out of speed? **5 s**

7. So what will be the arrow's speed 5 seconds after you shoot it? **0 m/s**

> Speed of free fall = acceleration × time = $10 \text{ m/s}^2 t$.

Average Speed

8. The arrow's initial upward speed is 50 m/s, and 5 seconds later it's zero. What's the average speed of the arrow during this time? **25 m/s**

9. What is the average speed of the arrow on the way back down— that is, for a beginning speed of zero and final speed of 50 m/s? **25 m/s**

> Average speed = $\dfrac{\text{initial speed + final speed}}{2}$

Free Fall Distance

10. Speed is one thing; distance another. How high will the arrow be 5 seconds after being shot up at 50 m/s? **125 m**

11. If an apple falls from a tree and hits the ground below in 1 second, its speed is 10 m/s when it hits. In words, explain why it falls a vertical distance of 5 meters, and NOT 10 meters.

> Distance = average speed × time.

DISTANCE OF FALL IS AVERAGE SPEED × TIME, ITS AVERAGE SPEED OF FALL IS 5 m/s.

7

CONCEPTUAL PHYSICAL SCIENCE EXPLORATIONS
Chapter 3 Newton's Second Law—Force and Acceleration
Free Fall

A rock dropped from the top of a cliff picks up speed as it falls. Pretend that a speedometer and odometer are attached to the rock to show readings of speed and distance at 1-second intervals. Both speed and distance are zero at time = zero (see sketch). Note that after falling 1 second the speed reading is 10 m/s and the distance fallen is 5 m. The readings for succeeding seconds of fall are not shown and are left for you to complete. So draw the position of the speedometer pointer and write in the correct odometer reading for each time. Use g = 10 m/s² and neglect air resistance.

> **YOU NEED TO KNOW:**
> Instantaneous speed of fall from rest:
> $$v = gt$$
> Average speed $\bar{v} = \dfrac{}{}$ final speed
> Distance fallen from rest:
> $$d = \bar{v}t$$

1. The speedometer reading increases by the same amount, **10** m/s, each second.

   This increase in speed per second is called **ACCELERATION**.

2. The distance fallen increases as the square of the **TIME**.

3. If it takes 7 seconds to reach the ground, then its speed at impact is **70** m/s, the total distance fallen is **245** m, and its acceleration of fall just before impact is **10** m/s².

Speedometer / odometer readings:
- t = 0 s: 0 0 0 meters
- t = 1 s: 0 0 5 meters
- t = 2 s: 2 0 meters
- t = 3 s: 4 5 meters
- t = 4 s: 8 0 meters
- t = 5 s: 1 2 5 meters
- t = 6 s: 1 8 0 meters

9

CONCEPTUAL PHYSICAL SCIENCE EXPLORATIONS

Chapter 3 Newton's Second Law—Force and Acceleration
Friction Force and Acceleration

Here's the crate filled with delicious peaches discussed in Chapter 4. The only vertical forces acting on the crate are gravity and the support force of the floor. The vector F_W is the weight, and F_N is the support force. We use the subscript N to indicate the force is normal (right angles) to the floor.

1. When Harry pulls with force P, 150 N, and the crate doesn't move, what is the magnitude of the friction? **150 N**

2. If the pulling force P is increased to 200 N and the crate doesn't move, what is the magnitude of f? **200 N**

3. So when Harry pulls with 200 N, the crate is in (static equilibrium) (dynamic equilibrium).

4. Harry increases his pull and the crate begins to slide. When he pulls with 250 N it slides at constant velocity. What is the magnitude of f? **250 N**

5. When Harry pulls with 250 N the crate is in (static equilibrium) (dynamic equilibrium).

6. If the mass of the crate is 50 kg and sliding friction is 250 N, what is the acceleration of the crate when pulled with 250 N? **0** 300 N? **1 m/s²** 500 N? **5 m/s²**

Force and Acceleration

1. Skelly the skater, total mass 25 kg, is propelled by rocket power.

a. Complete Table I
(neglect resistance)

$$a = \frac{F}{25\,kg}$$

TABLE I

FORCE	ACCELERATION
100 N	4 m/s²
200 N	8 m/s²
250 N	10 m/s²

b. Complete Table II for a constant 50-N resistance.

$$a = \frac{F - 50\,N}{25\,kg}$$

TABLE II

FORCE	ACCELERATION
50 N	0 m/s²
100 N	2 m/s²
200 N	6 m/s²

11

Chapter 3, or Appendix B: (Motion)
Hang Time

Some athletes and dancers have great jumping ability. When they leap straight up, they seem to momentarily "hang in the air" and defy gravity. The time that a jumper is airborne with feet off the ground is called hang time. Ask your friends to estimate the hang time of the great jumpers. They may say two or three seconds. But surprisingly, the hang time of the greatest jumpers is most always less than 1 second! A longer time is one of many illusions we have about nature.

To better understand this, find the answers to the following questions:

1. If you step off a table and it takes one-half second to reach the floor, what will be your speed when you meet the floor?

$$v = gt = 10 \tfrac{m}{s^2} \times \tfrac{1}{2}\,s = 5 \tfrac{m}{s}$$

Speed of free fall = acceleration × time
= 10 m/s² × number of seconds
= 10 t m.

2. What will be your average speed of fall?

$$\bar{v} = \frac{0 + 5\,m/s}{2} = 2.5 \tfrac{m}{s}$$

Average speed = $\dfrac{\text{initial speed} + \text{final speed}}{2}$

3. What will be the distance of fall?

$$d = \bar{v}t = 2.5\,m/s \times \tfrac{1}{2}\,s = 1.25\,m$$

Distance = average speed × time.

4. So how high is the surface of the table above the floor?

1.25 m

Jumping ability is best measured by a standing vertical jump. Stand facing a wall with feet flat on the floor and arms extended upward. Make a mark on the wall at the top of your reach. Then make your jump and at the peak make another mark. The distance between these two marks measures your vertical leap. If it's more than 0.6 meters (2 feet), you're exceptional.

5. What is your vertical jumping distance?

(VARIES)

6. Calculate your personal hang time. (Use the formula $d = 1/2\,gt^2$, from Appendix B. Remember that hang time is the time that you move upward + the time you return downward.)

Almost anybody can safely step off a 1.25-m (4-feet) high table. Can anybody in your school jump from the floor up onto the same table?

No way!

$$t = 2\sqrt{\frac{2d}{g}}$$

There's a big difference in how high you can reach and how high you raise your "center of gravity" when you jump. Even basketball star Michael Jordan in his prime couldn't quite raise his body 1.25 meters high, although he could easily reach higher than the more-than-3-meter high basket.

Here we're talking about vertical motion. How about running jumps? We'll see in Chapter 8 that the height of a jump depends only on the jumper's vertical speed at launch. While airborne, the jumper's horizontal speed remains constant while the vertical speed undergoes acceleration due to gravity. While airborne, no amount of leg or arm pumping or other bodily motions can change your hang time.

Fascinating physical science!

10

157

CONCEPTUAL PHYSICAL SCIENCE **EXPLORATIONS**

Chapter 4 Newton's Third Law—Action and Reaction

Action and Reaction Pairs

1. In the example below, the action-reaction pair is shown by the arrows (vectors), and the action-reaction described in words. In (a) through (g) draw the other arrow (vector) and state the reaction to the given action. Then make up your own example in (h).

Example:

Fist hits wall.
Wall hits fist.

(a) **Ball bumps head**
Head bumps ball.

(b) **Bug hits windshield**
Windshield hits bug.

(c) **Ball hits bat**
Bat hits ball.

(d) **Nose touches hand**
Hand touches nose.

(e) **Flower pulls on hand**
Hand pulls on flower.

(f) **Bar pushes athlete upward**
Athlete pushes bar upward.

(g) **Balloon surface pushes compressed air inward**
Compressed air pushes balloon surface outward.

(h) **Thing A acts on Thing B**
Thing B acts on Thing A

YOU CAN'T TOUCH WITHOUT BEING TOUCHED—NEWTON'S THIRD LAW

2. Draw arrows to show the chain of at least six pairs of action-reaction forces below.

13

Chapter 3 Newton's Second Law—Force and Acceleration
Mass and Weight

PHYSICS PHYSICS PHYSICS PHYSICS

1. Use the words *mass*, *weight*, and *volume*, to complete the table.

The force due to gravity on an object	WEIGHT
The quantity of matter in an object	MASS
The amount of space an object occupies	VOLUME

2. Different masses are hung on a spring scale calibrated in newtons.

⟵ 9.8 N

⟵ 1 kg

The force exerted by gravity on 1 kg = 9.8 N.
The force exerted by gravity on 5 kg = __49__ N.
The force exerted by gravity on __10__ kg = 98 N.

Make up your own mass and show the corresponding weight:
The force exerted by gravity on ___ kg = ___ N.

✱ **Any value for kg as long as the same value is multiplied by 9.8 for N.**

3. By whatever means (spring scales, measuring balance, etc.), find the mass of your physical science book. Then complete Table 1.

Table I

OBJECT	MASS	WEIGHT
MELON	1 kg	9.8 N
APPLE	0.1 kg	1 N
BOOK	1.3 kg	12.7 N
UNCLE HARRY	90 kg	882 N

4. Why isn't the girl hurt when the nail is driven into the block of wood? **The inertia of the block plays the main role; its tendency to remain at rest prevents it from sudden motion downward against her head.** Would this be more dangerous or less dangerous if the block were less massive __MORE__? Explain. **Less inertia means more downward motion of the block when struck.**

CAUTION: Safety dictates you not try this experiment yourself.

12

CONCEPTUAL PHYSICAL SCIENCE EXPLORATIONS

Chapter 5 Momentum
Changing Momentum

1. A moving car has momentum. If it moves twice as fast, its momentum is __TWICE__ as much.

2. Two cars, one twice as heavy as the other, move down a hill at the same speed. Compared with the lighter car, the momentum of the heavier car is __TWICE__ as much.

3. The recoil momentum of a cannon that kicks is
(more than) (less than) (the same as) the momentum of the cannonball it fires.

(Here we neglect the momentum of the gases.)

4. Suppose you are traveling in a bus at highway speed on a nice summer day and the momentum of an unlucky bug is suddenly changed as it splatters onto the front window.

a. Compared to the force that acts on the bug, how much force acts on the bus?

(more) (the same) (less)

b. The time of impact is the same for both the bug and the bus. Compared to the impulse on the bug, this means the impulse on the bus is

(more) (the same) (less)

c. Although the momentum of the bus is very large compared to the momentum of the bug, the change in momentum of the bus, compared to the change of momentum of the bug is

(more) (the same) (less)

d. Which undergoes the greater acceleration?

(bus) (both the same) (bug)

e. Which therefore, suffers the greater damage?

(bus) (both the same) (the bug of course)

Chapter 5 Momentum

5. Granny whizzes around the rink and is suddenly confronted with Ambrose at rest directly in her path. Rather than knock him over, she picks him up and continues in motion without "braking."

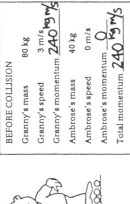

Consider both Granny and Ambrose as two parts of one system. Since no outside forces act on the system, the momentum of the system before collision equals the momentum of the system after collision.

a. Complete the before-collision data in the table below.

BEFORE COLLISION	
Granny's mass	80 kg
Granny's speed	3 m/s
Granny's momentum	__240 kg m/s__
Ambrose's mass	40 kg
Ambrose's speed	0 m/s
Ambrose's momentum	__0__
Total momentum	__240 kg m/s__

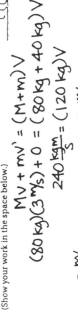

b. After collision, does Granny's speed increase or decrease?

__DECREASE__

c. After collision, does Ambrose's speed increase or decrease?

__INCREASE__

d. After collision, what is the total mass of Granny + Ambrose?

__120 kg__

e. After collision, what is the total momentum of Granny + Ambrose?

__240 kg m/s__

f. Use the conservation of momentum law to find the speed of Granny and Ambrose together after collision. (Show your work in the space below.)

$$Mv + mv' = (M+m)V$$
$$(80 kg)(3 m/s) + 0 = (80 kg + 40 kg)V$$
$$\frac{240 \frac{kg m}{s}}{} = (120 kg)V$$
$$V = 2 m/s$$

New speed = __2 m/s__

CONCEPTUAL PHYSICAL SCIENCE EXPLORATIONS

Chapter 5 Momentum
Systems

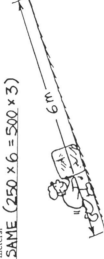

1. When the compressed spring is released, Blocks A and B will slide apart. There are 3 systems to consider here, indicated by the closed dashed lines below — System A, System B, and System A+B. Ignore the vertical forces of gravity and the support force of the table.

a. Does an external force act on System A? (yes) (no)
Will the momentum of System A change? (yes) (no)

b. Does an external force act on System B? (yes) (no)
Will the momentum of System B change? (yes) (no)

c. Does an external force act on System A+B? (yes) (no)
Will the momentum of System A+B change? (yes) (no)

Note that external forces on System A and System B are internal to System A+B so they cancel!!

2. Billiard ball A collides with billiard ball B at rest. Isolate each system with a closed dashed line. Draw only the external force vectors that act on each system.

a. Upon collision, the momentum of System A (increases) (decreases) (remains unchanged).
b. Upon collision, the momentum of System B (increases) (decreases) (remains unchanged).
c. Upon collision, the momentum of System A+B (increases) (decreases) (remains unchanged).

3. A girl jumps upward. In the sketch here, draw a closed dashed line to indicate the system of the girl.

a. Is there an external force acting on her? (yes) (no)
Does her momentum change? (yes) (no)
Is the girl's momentum conserved? (yes) (no)

b. In the sketch to the right, draw a closed dashed line to indicate the system [girl + earth]. Is there an external force due to the interaction between the girl and the earth that acts on the system? (yes) (no)
Is the momentum of the system conserved? (yes) (no)

4. A block strikes a blob of jelly. Isolate 3 systems with a closed dashed line and show the external force on each. In which system is momentum conserved?
ONE ON RIGHT

5. A truck crashes into a wall. Isolate 3 systems with a closed dashed line and show the the external force on each. In which system is momentum conserved?
ONE ON RIGHT

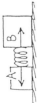

17

CONCEPTUAL PHYSICAL SCIENCE EXPLORATIONS

Chapter 6 Work and Energy
Work, Power, and Energy

1. How much work (energy) is needed to lift an object that weighs 200 N to a height of 4 m?
800 J

2. How much power is needed to lift the 200-N object to a height of 4 m in 4 s?
200 W

3. What is the power output of an engine that does 60,000 J of work in 10 s?
6000 W

4. The block of ice weighs 500 newtons.

a. Neglecting friction, how much force is needed to push it up the incline?
250 N

b. How much work is required to push it up 3 meters?
SAME (250 × 6 = 500 × 3)

5. All the ramps are 5 m high. We know that the KE of the block at the bottom of the ramp will be equal to the loss of PE (conservation of energy). Find the speed of the block at ground level in each case. [Hint: Do you recall from earlier chapters how long it takes something to fall a vertical distance of 5 m from a positon of rest (assume $g = 10$ m/s²)? And how much speed a falling object acquires in this time? This gives you the answer to Case 1. Cases 2 and 3 look more complex. But not so if you use energy conservation to guide your answer (no further computations needed). Discuss with your classmates how energy conservation gives you the answers to Cases 2 and 3.]

Case 1: Speed = **10** m/s   Case 2: Speed = **10** m/s   Case 3: Speed = **10** m/s.

SPEED SAME BECAUSE ΔKE SAME; BUT TIME IS DIFFERENT

19

CONCEPTUAL PHYSICAL SCIENCE **EXPLORATIONS**

Chapter 6 Work and Energy
Conservation of Energy

Fill in the blanks for the six systems shown.

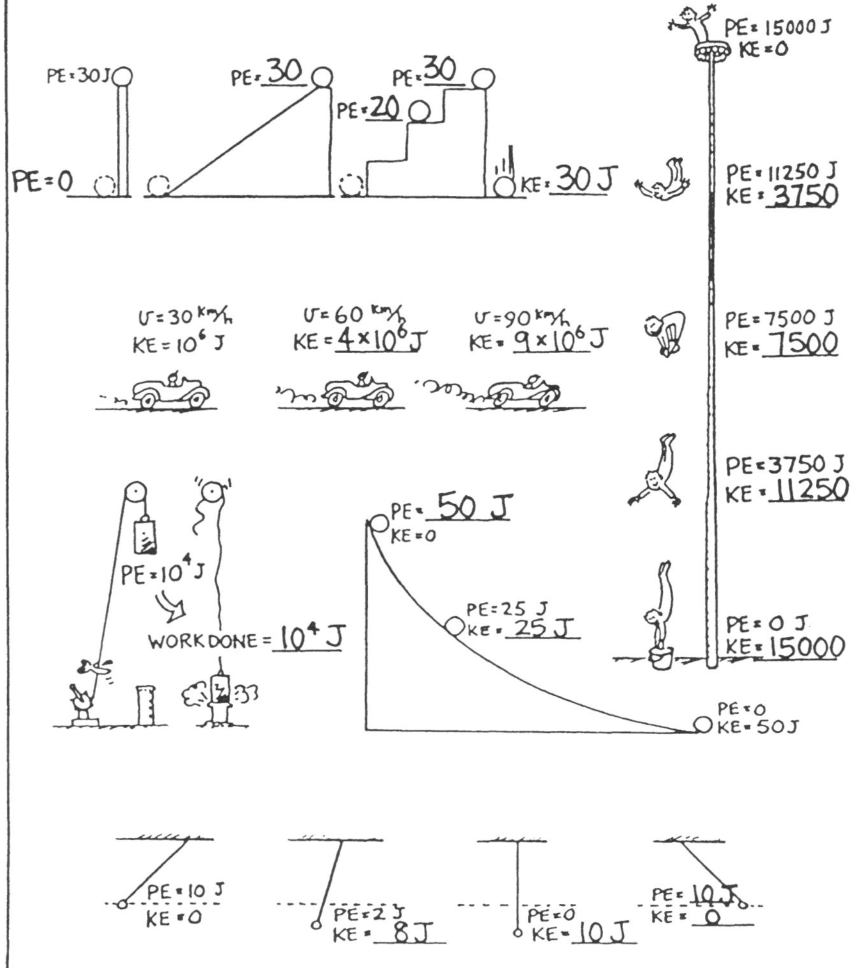

PE=30J KE=0

PE=30 PE=20 PE=30 KE=30 J

PE=0 KE=___

PE=75J KE=25 J   PE=25 J KE=75 J

v=30 km/h KE=10⁶ J

v=60 km/h KE=4×10⁶ J

v=90 km/h KE=9×10⁶ J

PE=10⁴ J   WORK DONE = 10⁴ J

PE=0 KE=50 J   PE=50 J

PE=25 J KE=25 J

PE=15000 J KE=0

PE=11250 J KE=3750

PE=7500 J KE=1500

PE=3750 J KE=11250

PE=0 J KE=15000

PE=0 KE=50J

PE=10 J KE=0

PE=2 J KE=8 J

PE=0 KE=10 J

PE=10 J KE=0

---

6. Which block gets to the bottom of the incline first? Assume no friction. (Be careful!) Explain your answer.

   BLOCK A. BECAUSE GREATER ACCELERATION AND LESS RAMP DISTANCE... SO A HAS SHORTER SLIDING TIME... BUT SAME SPEED

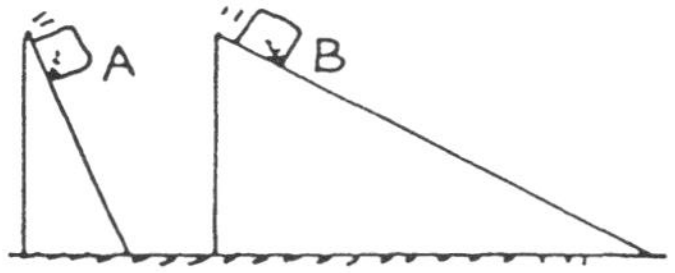

   A   B

7. The KE and PE of a block freely sliding down a ramp are shown in only one place in the sketch. Fill in the missing values.

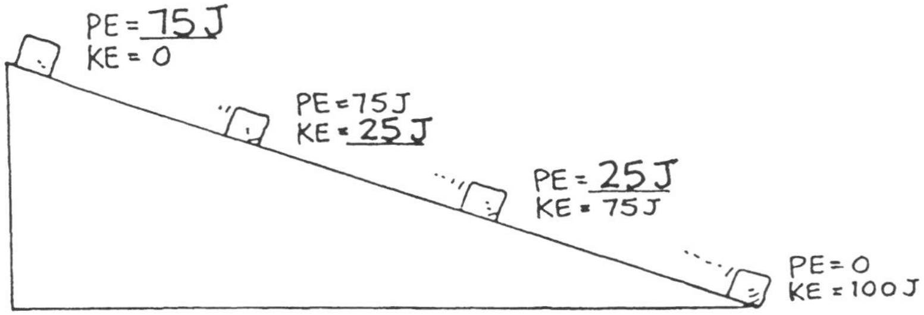

   PE=75J KE=25 J
   PE=50 KE=50
   PE=25 J KE=75 J
   PE=0 KE=100 J

8. A big metal bead slides due to gravity along an upright friction-free wire. It starts from rest at the top of the wire as shown in the sketch. How fast is it traveling as it passes

   Point B? 10 m/s
   Point D? 10 m/s
   Point E? 10 m/s

   At what point does it have the maximum speed? C

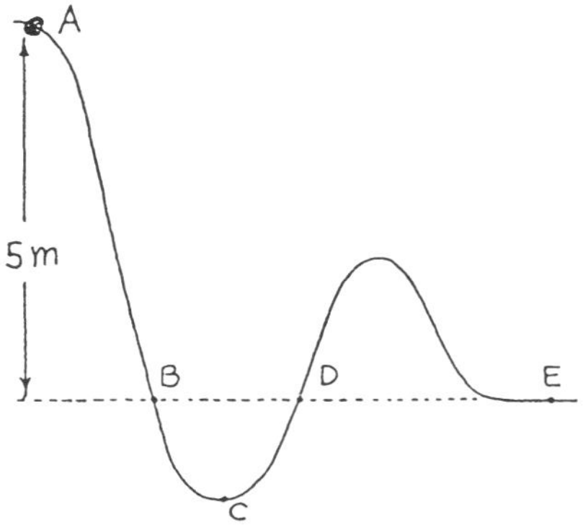

   5 m

   A B C D E

9. Rows of wind-powered generators are used in various windy locations to generate electric power. Does the power generated affect the speed of the wind? Would locations behind the 'windmills' be windier if they weren't there? Discuss this in terms of energy conservation with your classmates.

   YES, BY CONS OF ENERGY, ENERGY GAINED BY WINDMILLS IS TAKEN FROM WIND KE. SO WIND MUST SLOW DOWN. LOCATIONS BEHIND WOULD BE A BIT WINDIER WITHOUT THE WINDMILLS!

CONCEPTUAL PHYSICAL SCIENCE EXPLORATIONS

Chapter 7 Gravity
Inverse-Square Law

1. Paint spray travels radially **away** from the nozzle of the can in straight lines. Like gravity, the strength (intensity) of the spray obeys an inverse-square law. Complete the diagram by filling in the blank spaces.

1m 2m 3m
4m

PAINT GUN

PAINT SPRAY	1 AREA UNIT	4 AREA UNITS	(9) AREA UNITS	(16) AREA UNITS
	1 mm THICK	¼ mm THICK	(⅑) mm THICK	(1/16) mm THICK

2. A small light source located 1 m in front of an opening of area 1 m² illuminates a wall behind. If the wall is 1 m behind the opening (2 m from the light source), the illuminated area covers 4 m². How many square meters will be illuminated if the wall is

1 m² OPENING

LIGHT SOURCE

4 m² OF ILLUMINATION

1 m — 2 m

5 m from the source? __25 m²__

10 m from the source? __100 m²__

3. The sound of a bullfrog attracts a potential mate. How much quieter is the sound if the potential mate is 5 times farther away?

__⅕ AS LOUD, so 25 TIMES QUIETER__

4. The sound of a cricket gets stronger as you get closer to it. If you get twice as close, how much louder will the sound be?

__4 TIMES AS LOUD__

5. You are on a spaceship approaching the planet Jupiter. You feel its gravitational pull. When you get three times closer to Jupiter, how much stronger will the pull?

__9 TIMES AS STRONG__

6. At a certain distance from each other, two asteroids in space feel a gravitational attraction to each other of 1000 N. When their distance apart is reduced to half the distance, what will be the gravitational force on each asteroid?

__250 N__

Chapter 6 Work and Energy
Machines

1. The woman supports a 100-N load with the friction-free pulley systems shown below. Fill in the spring-scale readings that show how much force she must exert.

100 N

50 N

50 N

2. A 600-N block is lifted by the friction-free pulley system shown.

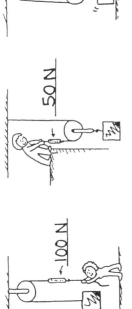

600 N

a. How many strands of rope support the 600-N weight?

__6__

b. What is the tension in each strand?

__100 N__

c. What is the tension in the end held by the man?

__100 N__

d. If the man pulls his end down 60 cm, how many cm will the weight rise?

__10 cm__

e. Does the man multiply force or energy (or both)?

__ONLY FORCE ! MULTIPLYING ENERGY IS IMPOSSIBLE.__

f. If the man does 60 joules of work, what will be the increase of PE of the 600-N weight?

__60 J__

3. Why don't balls bounce as high during the second bounce as they do in the first?

__DURING EACH BOUNCE SOME OF BALL'S MECHANICAL ENERGY IS TRANSFORMED INTO HEAT (AND EVEN SOUND), SO PE DECREASES WITH EACH BOUNCE.__

Chapter 7 Gravity
Force and Weight

1. An apple that has a mass of 0.1 kilogram has the same mass wherever it is. The amount of matter that makes up the apple (depends upon) (does not depend upon) the location of the apple. It has the same resistance to acceleration wherever it is — its inertia everywhere is (the same) (different).

The weight of the apple is a different story. It may weigh exactly 1 N in San Francisco and slightly less in mile-high Denver, Colorado. On the surface of the moon the apple would weigh 1/6 N, and far out in outer space it may have almost no weight at all. The quantity that doesn't change with location is (mass) (weight),

and the quantity that may change with location is its (mass) (weight).

That's because (mass) (weight)

is the force due to gravity on a body, and this force varies with distance. So weight is the force of gravity between two bodies, usually some small object in contact with the Earth. When we refer to the (mass) (weight)

of an object we are usually speaking of the gravitational force that attracts it to the Earth.

Fill in the blanks:

2. If we stand on a weighing scale and find that we are pulled toward the Earth with a force of 500 N, then we weigh **500** N. Strictly speaking, we weigh **500** N relative to the Earth. How much does the Earth weigh? If we tip the scale upside down and repeat the weighing process, we can say that we and the Earth are still pulled together with a force of **500** N, and therefore, relative to us, the whole 6,000,000,000,000,000,000,000,000-kg Earth weighs **500** N! Weight, unlike mass, is a relative quantity.

VIEW THE SAME FROM ANOTHER PERSPECTIVE!

DO YOU SEE WHY IT MAKES SENSE TO DISCUSS THE EARTH'S MASS, BUT NOT ITS WEIGHT?

We are pulled to the Earth with a force of 500 N, so we weigh 500 N.

The Earth is pulled toward us with a force of 500 N.

Chapter 7 Gravity
Our Ocean Tides

1. Consider two equal-mass blobs of water, A and B, initially at rest in the moon's gravitational field. The vector shows the gravitational force of the moon on A.

a. Draw a force vector on B due to the moon's gravity.

b. Is the force on B more or less than the force on A? **LESS**

c. Why? **FARTHER AWAY**

d. The blobs accelerate toward the moon. Which has the greater acceleration? (A) (B)

e. Because of the different accelerations, with time (A gets farther ahead of B) (A and B gain identical speeds) and the distance between A and B (increases) (stays the same) (decreases).

f. If A and B were connected by a rubber band, with time the rubber band would (stretch) (not stretch).

g. This (stretching) (nonstretching) is due to the (difference) (nondifference) in the moon's gravitational pulls.

h. The two blobs will eventually crash into the moon. To orbit around the moon instead of crashing into it, the blobs should move (away from the moon) (tangentially). Then their accelerations will consist of changes in (speed) (direction).

2. Now consider the same two blobs located on opposite sides of the Earth.

a. Because of differences in the moon's pull on the blobs, they tend to (spread away from each other) (approach each other).

b. Does this spreading produce ocean tides? (Yes) (No)

c. If Earth and moon were closer, gravitational force between them would be (more) (the same) (less), and the difference in gravitational forces on the near and far parts of the ocean would be (more) (the same) (less).

d. Because the Earth's orbit about the sun is slightly elliptical, Earth and sun are closer in December than in June. Taking the sun's tidal force into account, on a world average, ocean tides are greater in (December) (June) (no difference).

Chapter 8 Projectile and Satellite Motion
Projectiles

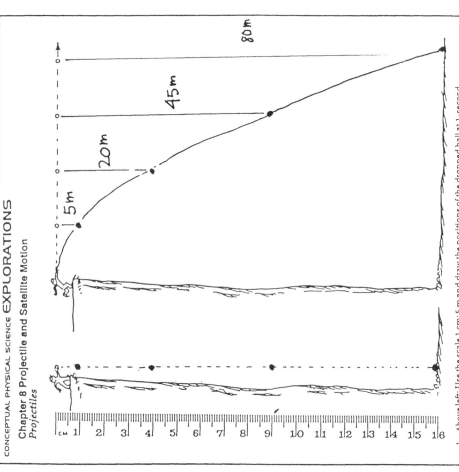

80 m

45 m

20 m

5 m

1. Above left: Use the scale 1 cm: 5 m and draw the positions of the dropped ball at 1-second intervals. Neglect air drag and assume $g = 10$ m/s². Estimate the number of seconds the ball is in the air. __4__ seconds.

2. Above right: The four positions of the thrown ball with *no gravity* are at 1-second intervals. At 1 cm: 5 m, carefully draw the positions of the ball *with* gravity. Neglect air drag and assume $g = 10$ m/s². Connect your positions with a smooth curve to show the path of the ball. How is the motion in the vertical direction affected by motion in the horizontal direction? ONLY VERT MOTION AFFECTED BY GRAVITY; HORIZ MOTION INDEPENDENT

27

80 m

45 m

20 m

5 m

3. This time the ball is thrown below the horizontal. Use the same scale 1 cm: 5 m and carefully draw the positions of the ball as it falls beneath the dashed line. Connect your positions with a smooth curve. Estimate the number of seconds the ball remains in the air. __3.5__ s

4. Suppose that you are an accident investigator and you are asked to figure whether or not the car was speeding before it crashed through the rail of the bridge and into the mudbank as shown. The speed limit on the bridge is 55 mph = 24 m/s. What is your conclusion?

Car covers 24 m in 1 sec (5 m drop), so going 24 m/s after crashing thru rail. So it must have been going faster before hitting rail. So driver was speeding!

28

CONCEPTUAL PHYSICAL SCIENCE **EXPLORATIONS**

Chapter 8 Projectile and Satellite Motion
Circular Orbit

1. Figure A shows "Newton's Mountain," so high that its top is above the drag of the atmosphere. The cannonball is fired and hits the ground as shown.

   a. You draw the path the cannonball might take if it were fired a little bit faster.

   b. Repeat for a still greater speed, but still less than 8 km/s.

   c. Then draw the orbital path it would take if its speed were 8 km/s.

   d. What is the shape of the 8 km/s curve? __CIRCLE__

   e. What would be the shape of the orbital path if the cannonball were fired at a speed of about 9 km/s? __ELLIPSE__

— Figure A

2. Figure B shows a satellite in circular orbit.

   a. At each of the four positions draw a vector that represents the gravitational force exerted on the satellite.

   b. Label the force vectors F.

   c. Then draw at each position a vector to represent the velocity of the satellite at that position, and label it V.

   d. Are all four F vectors the same length? Why or why not? __YES; SATELLITE IS AT SAME DISTANCE, SAME FORCE__

   e. Are all four V vectors the same length? Why or why not? __YES; IN CIRCULAR ORBIT F⊥V SO THERE'S NO COMPONENT OF FORCE ALONG U TO CHANGE SPEED U.__

   f. What is the angle between your F and V vectors? __90°__

   g. If an enormously tall bowling alley extended as high as the satellite orbits, can you see that the force of gravity wouldn't change the speed of the ball (because it's always pulling at right angles to the alley)? So what does this tell you about the work that the force of gravity does on a satellite in circular orbit? __NO WORK BECAUSE THERE'S NO COMP OF FORCE ALONG PATH__

   h. Does the kinetic energy of the satellite in Figure B remain constant, or does it vary? __CONSTANT__

   i. Does the potential energy of the satellite remain constant, or does it vary? __REMAINS CONSTANT__

Figure B

Chapter 8 Projectile and Satellite Motion
Elliptical Orbit

3. Figure C shows a satellite in elliptical orbit.

   a. Repeat the procedure you used for the circular orbit, drawing vectors F and V for each position, including proper labeling. Show equal magnitudes with equal lengths, and greater magnitudes with greater lengths, but don't bother making the scale accurate.

   b. Are your vectors F all the same magnitude? Why or why not? __NO, FORCE DECREASES WHEN DISTANCE FROM EARTH INCREASES__

   c. Are your vectors V all the same magnitude? Why or why not? __NO, WHEN KE DECREASES (AS SATELLITE MOVES FARTHER FROM EARTH) SPEED DECREASES. WHEN KE INCREASES (CLOSER TO EARTH) SPEED INCREASES__

   d. Is the angle between vectors F and V everywhere the same, or does it vary? __IT VARIES__

   e. Are there places where there is a component of F along V? (That is, are there places where the force of gravity is not at right angles to the orbit.) __YES (EVERYWHERE EXCEPT AT THE APOGEE AND PERIGEE)__

   f. Is work done on the satellite when there is a component of F along and in the same direction of V and if so, does this increase or decrease the KE of the satellite? __YES. THIS INCREASES KE OF SATELLITE__

   g. When there is a component of F along and opposite to the direction of V, does this increase or decrease the KE of the satellite? __THIS DECREASES KE OF SATELLITE__

   h. What can you say about the sum KE + PE along the orbit? __CONSTANT (IN ACCORD WITH CONSERVATION OF ENERGY)__

Figure C

Be very very careful when placing both velocity and force vectors on the same diagram. Not a good practice, for one may construct the resultant of the vectors -- ouch!

CONCEPTUAL PHYSICAL SCIENCE EXPLORATIONS

Chapter 9 Thermal Energy
Temperature, Heat, and Expansion

1. Complete the table:

TEMPERATURE OF MELTING ICE	0°C	32°F	273 K
TEMPERATURE OF BOILING WATER	100°C	212°F	373 K

2. Suppose you apply a flame and heat one liter of water, raising its temperature 10°C. If you transfer the same heat energy to two liters, how much will the temperature rise? For three liters? *Record your answers on the blanks in the drawing at the right.*

$\Delta T = 10°C \qquad \Delta T = 5 °C \qquad \Delta T = 3.3°C$

3. A thermometer is in a container half-filled with 20°C water.

   a. When an equal volume of 20°C water is added, the temperature of the mixture is

   (10°C) (20°C) (40°C)

   b. When instead an equal volume of 40°C water is added, the temperature of the mixture will be

   (20°C) (30°C) (40°C)

   c. When instead a small amount of 40°C water is added, the temperature of the mixture will be

   (20°C) (between 20°C and 30°C) (30°C) (more than 30°C)

4. A red-hot piece of iron is put into a bucket of cool water. *Mark the following statements true (T) or false (F).* (Ignore heat transfer to the bucket.)

   a. The decrease in iron temperature equals the increase in the water temperature. __F__

   b. The quantity of heat lost by the iron equals the quantity of heat gained by the water. __T__

   c. The iron and water both will reach the same temperature. __T__

   d. The final temperature of the iron and water is halfway between the initial temperatures of each. __F__

CAN COMMON ICE BE COLDER THAN 0°C? YES

31

Chapter 9 Thermal Energy
Thermal Expansion

1. The weight hangs above the floor from the copper wire. When a candle is moved along the wire and heats it, what happens to the height of the weight above the floor? Why?

   HEIGHT DECREASES AS WIRE LENGTHENS

2. The levels of water at 0°C and 1°C are shown below in the first two flasks. At these temperatures there is microscopic slush in the water. There is slightly more slush at 0°C than at 1°C. As the water is heated, some of the slush collapses as it melts, and the level of the water falls in the tube. That's why the level of water is slightly lower in the 1°C-tube. Make rough estimates and sketch in the appropriate levels of water at the other temperatures shown. What is important about the level when the water reaches 4°C?

   SINCE WATER IS MOST DENSE AT 4°C, WATER LEVEL IS LOWEST AT 4°C

   0°C   1°C   2°C   3°C   4°C   5°C   6°C

3. The diagram at right shows an ice-covered pond. Mark the probable temperatures of water at the top and bottom of the pond.

   ICE 0°C
   4°C

   WHICH WILL WEIGH MORE, 1 LITER OF ICE OR 1 LITER OF WATER? WATER (MORE DENSE)

   I CAN'T GET THIS METAL LID OFF THE JAR.... SHOULD I HEAT THE LID OR COOL IT? WHY? HEAT IT SO IT WILL EXPAND

32

CONCEPTUAL PHYSICAL SCIENCE EXPLORATIONS

Chapter 10 Heat Transfer and Change of Phase
Change of Phase

All matter can exist in the solid, liquid, or gaseous phases. The solid phase exists at relatively low temperatures, the liquid phase at higher temperatures, and the gaseous phase at still higher temperatures. Water is the most common example, not only because of its abundance but also because the temperatures for all three phases are common.

1. How many calories are needed to change 1 gram of 0°C ice to water?
$\boxed{\text{ICE } 0°C} \Rightarrow \boxed{\text{WATER } 0°C}$
80

2. How many calories are needed to change the temperature of 1 gram of water by 1°C?
$\boxed{T} \Rightarrow \boxed{T+1°C}$
1

3. How many calories are needed to melt 1 gram of 0°C ice and turn it to water at a room temperature of 23°C?
$\boxed{0°C} \Rightarrow \boxed{23°C}$
80 CAL + 23 CAL = 103 CAL

4. A 50-gram sample of ice at 0°C is placed in a glass beaker that contains 200 g of water at 20°C.
   a. How much heat is needed to melt the ice? **4000 CAL**
   SINCE THERE'S 50g OF ICE AND 80 CAL IS REQUIRED PER GRAM, HEAT REQUIRED IS 50g × (80 CAL/g) = 4000 CAL
   b. By how much would the temperature of the water change if it gave up this much heat to the ice? **BY 20°C**
   200 g OF WATER GIVES OFF 200 CAL FOR EACH 1°C DROP IN TEMP. SO 4000 CAL/200 CAL/°C = 20°C
   c. What will be the final temperature of the mixture? (Disregard any heat absorbed by the glass or given off by the surrounding air.) **0°C**

5. How many calories are needed to change 1 gram of 100°C boiling water to 100°C steam?
$\boxed{100°C} \Rightarrow 100°C$
540 CAL

6. Fill in the number of calories at each step below for changing the state of 1 gram of 0°C ice to 100°C steam.

$\boxed{\text{1 GRAM ICE 0°C}} \xrightarrow{\text{CHANGE OF PHASE}} \boxed{\text{1 GRAM WATER 0°C}} \xrightarrow{\text{TEMP. RISE}} \boxed{\text{1 GRAM WATER 100°C}} \xrightarrow{\text{CHANGE OF PHASE}} \boxed{\text{1 GRAM STEAM 100°C}}$

HEAT NEEDED = **80** CAL + **100** CAL + **540** CAL = **720** CAL

CONCEPTUAL PHYSICAL SCIENCE EXPLORATIONS

Chapter 10 Heat Transfer and Change of Phase
Evaporation

1. Why does it feel colder when you swim at a pool on a windy day?
Water evaporates from your body faster and cools you [*Energy is nature's way of keeping score!*]

2. Why does your skin feel cold when a little rubbing alcohol is applied to it?
Alcohol rapidly evaporates and cools you in process.

3. Briefly explain from a molecular point of view why evaporation is a cooling process.
The more energetic and faster molecules escape into the air. Energy taken with them reduces average KE of remaining molecules.

4. When hot water rapidly evaporates, the result can be dramatic. Consider 4 g of boiling water spread over a large surface so that 1 g rapidly evaporates. Suppose further that the surface and surroundings are very cold so that all 540 calories for evaporation come from the remaining 3 g of water.
   a. How many calories are taken from each gram of water?
   $\dfrac{540 \text{ CAL}}{3} = 180$ CAL
   b. How many calories are released when 1 g of 100°C water cools to 0°C?
   100 CAL
   c. How many calories are released when 1 g of 0°C water changes to 0°C ice?
   80 CAL
   d. What happens in this case to the remaining 3 g of boiling water when 1 g rapidly evaporates?
   The remaining water freezes! (Each gram of water releases 180 calories in cooling and freezing)

CONCEPTUAL PHYSICAL SCIENCE **EXPLORATIONS**

Chapter 11 Electricity
Coulomb's Law

1. The diagram is of a hydrogen atom.

   a. Label the proton in the nucleus with a + sign and the orbital electron with a - sign.

   b. The electrical interaction between the nucleus and the orbital electron is a force of

   (attraction) (repulsion)

   c. According to Coulomb's law,

   $$F = k\frac{q_1 q_2}{d^2}$$

   if the charge of either the nucleus or the orbital electron were greater, the force between the nucleus and the electron would be

   (greater) (less)

   and if the distance between the nucleus and electron were greater the force would be

   (greater) (less).

   If the distance between the nucleus and electron were doubled, the force would be

   (1/4 as much) (1/2 as much) (two times as much) (4 times as much)

2. Consider the electric force between a pair of charged particles a certain distance apart. By Coulomb's law.

   a. If the charge on one of the particles is doubled, the force is

   (unchanged) (halved) (doubled) (quadrupled)

   b. If, instead, the charge on both particles is doubled, the force is

   (unchanged) (halved) (doubled) (quadrupled)

   c. If instead the distance between the particles is halved, the force is

   (unchanged) (halved) (doubled) (quadrupled)

   d. If the distance is halved, and the charge of both particles is doubled, the force is   16   times as great.

$$F = k\frac{q_1 \cdot q_2}{d^2}$$

PHYSICS PHYSICS

CONCEPTUAL PHYSICAL SCIENCE **EXPLORATIONS**

Chapter 11 Electricity
Electric Pressure—Voltage

1. Just as PE (potential energy) transforms to KE (kinetic energy) for a mass lifted against the gravitational field (left), the electric PE of an electrically-charged particle transforms to other forms of energy when it changes location in an electric field (right). When released, how does the KE acquired by a charged particle compare with the decrease in PE?

   SAME

   PE

   KE

   PE +-

   KE +-

2. *Complete the statements.*

   A force compresses the spring. The work done in compression is the product of the average force and the distance moved. W = Fd. This work increases the PE of the spring.

   Similarly, a force pushes the charge (call it a test charge) closer to the charged sphere. The work done in moving the test charge is the product of the average **FORCE** and the **DISTANCE** moved. W = **Fd**. This work **INCREASES** the PE of the test charge.

   If the test charge is released, it will be repelled and fly past the starting point. Its gain in KE at this point is **EQUAL** to its decrease in PE.

   At any point, a greater quantity of test charge means a greater amount of PE, but not a greater amount of PE *per quantity of charge*. The quantities PE (measured in joules) and PE/charge (measured in volts) are different concepts.

   By definition: Voltage = PE/charge. 1 volt = 1 joule/1 coulomb.

3. *Complete the statements.*

   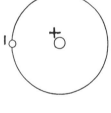

   ELECTRIC PE/CHARGE HAS THE SPECIAL NAME **VOLTAGE**

   VOLTAGE IS A SORT OF "ELECTRIC PRESSURE".

CONCEPTUAL PHYSICAL SCIENCE **EXPLORATIONS**

Chapter 11 Electricity

Ohm's Law

CURRENT = $\dfrac{\text{VOLTAGE}}{\text{RESISTANCE}}$ OR $I = \dfrac{V}{R}$

USE OHM'S LAW IN THE TRIANGLE TO FIND THE QUANTITY YOU WANT, COVER THE LETTER WITH YOUR FINGER AND THE REMAINING TWO SHOW YOU THE FORMULA?

CONDUCTORS AND RESISTORS HAVE RESISTANCE TO THE CURRENT IN THEM.

OHM MY GOODNESS!

1. How much current flows in a 1000-ohm resistor when 1.5 volts are impressed across it?
   __0.0015 A__

2. If the filament resistance in an automobile headlamp is 3 ohms, how many amps does it draw when connected to a 12-volt battery?
   __4 A__

3. The resistance of the side lights on an automobile are 10 ohms. How much current flows in them when connected to 12 volts?
   __1.2 A__

4. What is the current in the 30-ohm heating coil of a coffee maker that operates on a 120-volt circuit?
   __4 A__

5. During a lie detector test, a voltage of 6 V is impressed across two fingers. When a certain question is asked, the resistance between the fingers drops from 400,000 ohms to 200,000 ohms. What is the current (a) initially through the fingers, and (b) when the resistance between them drops?
   (a) __0.000015 A (15 μA)__ (b) __0.000030 A (30 μA)__

6. How much resistance allows an impressed voltage of 6 V to produce a current of 0.006 A?
   __1000 Ω__

7. What is the resistance of a clothes iron that draws a current of 12 A at 120 V?
   __10 Ω__

8. What is the voltage across a 100-ohm circuit element that draws a current of 1 A?
   __100 V__

9. What voltage will produce 3 A through a 15-ohm resistor?
   __45 V__

10. The current in an incandescent lamp is 0.5 A when connected to a 120-V circuit, and 0.2 A when connected to a 10-V source. Does the resistance of the lamp change in these cases? Explain your answer and defend it with numerical values.
    __YES, RESISTANCE INCREASES WITH HIGHER TEMP OF GREATER CURRENT.__
    __AT 0.2A, R = $\frac{10V}{3A}$ = 50Ω; AT 0.5A, R = $\frac{120V}{0.5A}$ = 240Ω__
    __(APPRECIABLY GREATER)__

---

4. When a charge of 1 C in an electric field has an electric PE of 1 J, it has a voltage of 1 V. When a charge of 2 C has an electric PE of 2 J, its voltage is = __1__ V.

$\dfrac{1J}{1C} = 1V$ PHYSICS $\dfrac{2J}{2C} = \ldots$

5. If a conductor connected to the terminal of a battery has a potential of 12 volts, then each coulomb of charge on the conductor has a PE of __12__ J.

6. If a charge of 1 C has a PE of 5000 J, its voltage is __5000__ V.

VOLTAGE = $\dfrac{PE}{CHARGE}$ = $\dfrac{0.5J}{0.0001C}$ = $\ldots$

7. If a charge of 0.001 C has a PE of 5 J, its voltage is __5000__ V.

8. If a charge of 0.0001 C has a PE of 0.5 J, its voltage is __5000__ V.

5000 volts?

9. If a rubber balloon is charged to 5000 V, and the quantity of charge on the balloon is 1 millionth coulomb, (0.000001 C) then the PE of this charge is only __0.005__ J.

10. Some people get mixed up between force and pressure. Recall that pressure is force *per area*. Similarly, some people get mixed up between electric PE and voltage. According to this chapter, voltage is electric PE *per* __CHARGE__.

Experiment, not philosophical discussion, decides what is correct in science.

YEA!

CONCEPTUAL PHYSICAL SCIENCE EXPLORATIONS

Chapter 11 Electricity
Series Circuits

1. In the circuit shown at the right, a voltage of 6 V pushes charge through a single resistor of 2 Ω. According to Ohm's law, the current in the resistor (and therefore in the whole circuit) is ____**3**____ A.

2Ω

6V

2. If a second identical lamp is added, as on the left, the 6-V battery must push charge through a total resistance of __**4**__ Ω. The current in the circuit is then **1.5** A.

2Ω 2Ω

6V

THE EQUIVALENT RESISTANCE OF RESISTORS IN SERIES IS SIMPLY THEIR SUM?

3. The equivalent resistance of three 4-Ω resistors in series is ____**12**____ Ω.

4. Does current flow *through* a resistor, or *across* a resistor? **THROUGH** Is voltage established *through* a resistor, or *across* a resistor? **ACROSS**

5. Does current in the lamps occur simultaneously, or does charge flow first through one lamp, then the other, and finally the last in turn? **SIMULTANEOUSLY (SPEED OF LIGHT)**

6. Circuits a and b below are identical with all bulbs rated at equal wattage (therefore equal resistance). The only difference between the circuits is that Bulb 5 has a short circuit, as shown.

a   1 2 3   b   4 5 6

4.5 V   4.5 V

   a. In which circuit is the current greater? **b**

   b. In which circuit are all three bulbs equally bright? **a**

   c. What bulbs are the brightest? **4 AND 6**

   d. What bulb is the dimmest? **5 (NOT LIT)**

   e. What bulbs have the largest voltage drops across them? **4 AND 6 (2.25V EACH)**

   f. Which circuit dissipates more power? **b (GREATER CURRENT, SAME VOLTAGE)**

   g. What circuit produces more light? **b (MORE POWER)**

Chapter 11 Electricity
Electric Power

The rate that energy is converted from one form to another is *power*.

$$power = \frac{energy\ converted}{time} = \frac{voltage \times charge}{time} = voltage \times \frac{charge}{time} = voltage \times current$$

The unit of power is the *watt* (or *kilowatt*). So in units form,

  Electric power (*watts*) = current (*amperes*) x voltage (*volts*),

where 1 *watt* = 1 *ampere* x 1 *volt*.

THAT'S RIGHT... VOLTAGE × $\frac{ENERGY}{CHARGE}$, SO ENERGY = VOLTAGE × CHARGE... AND $\frac{CHARGE}{TIME}$ = CURRENT ⸫ HEAT

A 100-WATT BULB CONVERTS ELECTRIC ENERGY INTO HEAT AND LIGHT MORE QUICKLY THAN A 25-WATT BULB. THAT'S WHY FOR THE SAME VOLTAGE A 100-WATT BULB GLOWS BRIGHTER THAN A 25-WATT BULB!

1. What is the power when a voltage of 120 V drives a 2-A current through a device? **240 W**

WHICH DRAWS MORE CURRENT ... THE 100-WATT OR THE 25-WATT BULB?

2. What is the current when a 60-W lamp is connected to 120 V? **0.5 A**

3. How much current does a 100-W lamp draw when connected to 120 V? **0.83 A**

4. If part of an electric circuit dissipates energy at 6 W when it draws a current of 3 A, what voltage is impressed across it? **2 V**

WATT'S HAPPENING?

5. The equation

$$power = \frac{energy\ converted}{time}$$

rearranged gives

  energy converted = **POWER × TIME**

6. Explain the difference between a kilowatt and a kilowatt-hour. **A KILOWATT IS A UNIT OF POWER; A KILOWATT-HOUR IS A UNIT OF ENERGY (POWER × TIME).**

7. For safety reasons, some people leave their front porch light on all the time. If they use a 60-W bulb at 120 V, and the local power utility sells energy at 15 cents per kilowatt-hour, how much does it cost to leave the bulb on for a month? Show your work on the other side of this page.

$$E = P \times t = 60W \times 1\,mo \times \frac{30d}{1\,mo} \times \frac{24h}{1d} \times \frac{1\,kW}{1000W} = 43.2\ kWh$$

$$cost = E \times rate = 43.2\ kWh \times \frac{\$0.15}{kWh} = \$6.48$$

CONCEPTUAL PHYSICAL SCIENCE EXPLORATIONS

Chapter 12 Magnetism

Magnetism

Fill in each blank with the appropriate word.

1. Attraction or repulsion of charges depends on their *signs*, positives or negatives. Attraction or repulsion of magnets depends on their magnetic __POLES__. Opposite poles __NORTH__ or __SOUTH__.

2. Opposite poles attract; like poles __REPEL__.

3. A magnetic field is produced by the __MOTION__ of electric charge.

4. Clusters of magnetically aligned atoms are magnetic __DOMAINS__.

5. A magnetic __FIELD__ surrounds a current-carrying wire.

6. When a current-carrying wire is made to form a coil around a piece of iron, the result is an __ELECTROMAGNET__.

7. A charged particle moving in a magnetic field experiences a deflecting that is maximum when the charge moves __PERPENDICULAR__ to the field.

8. A current-carrying wire experiences a deflecting __FORCE__ that is maximum when the wire and magnetic field are __PERPENDICULAR__ to one another.

9. A simple instrument designed to detect electric current is the __GALVANOMETER__; when calibrated to measure current, it is an __AMMETER__; when calibrated to measure voltage, it is a __VOLTMETER__.

10. The largest size magnet in the world is the __WORLD (OR EARTH)__ itself.

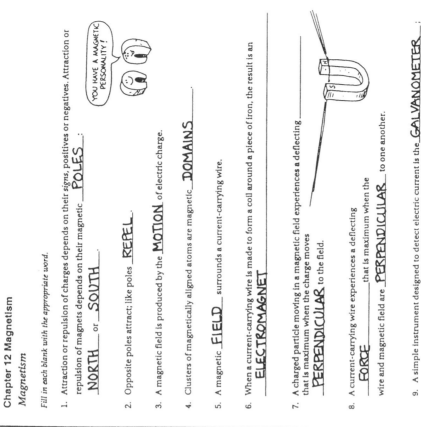

YOU HAVE A MAGNETIC PERSONALITY!

THEN TO REALLY MAKE THINGS "SIMPLE," THERE'S THE RIGHT-HAND RULE!

45

Chapter 11 Electricity

Parallel Circuits

1. In the circuit shown below, there is a voltage drop of 6 V across *each* 2-Ω resistor.

   a. By Ohm's law, the current in each resistor is __3__ A.

   b. The current through the battery is the sum of the currents in the resistors, __6__ A.

   c. Fill in the current in the eight blank spaces in the view of the *same circuit* shown again at the right.

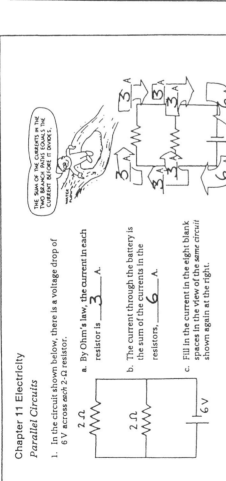

THE SUM OF THE CURRENTS IN THE TWO BRANCH PATHS EQUALS THE CURRENT BEFORE IT DIVIDES.

2. Cross out the circuit below that is not equivalent to the circuit above.

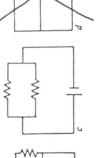 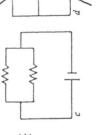 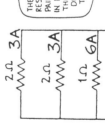

3. Consider the parallel circuit at the right.

   a. The voltage drop across each resistor is __6__ V.

   b. The current in each branch is:
      2-Ω resistor __3__ A
      2-Ω resistor __3__ A
      1-Ω resistor __6__ A

   c. The current through the battery equals the sum of the currents which equals __12__ A.

   d. The equivalent resistance of the circuit equals __0.5__ Ω.

THE EQUIVALENT RESISTANCE OF A PAIR OF RESISTORS IN PARALLEL IS THEIR PRODUCT DIVIDED BY THEIR SUM.

2Ω 3A
2Ω 3A
1Ω 6A
6V

44

CONCEPTUAL PHYSICAL SCIENCE **EXPLORATIONS**

Chapter 12 Magnetism

Faraday's Law

1. Hans Christian Oersted discovered that magnetism and electricity are (related) (independent of each other).

   Magnetism is produced by (batteries) (the motion of electric charges).

   Faraday and Henry discovered that electric current can be produced by (batteries) (motion of a magnet).

   More specifically, voltage is induced in a loop of wire if there is a change in the (batteries) (magnetic field in the loop).

   This phenomenon is called (electromagnetism) (electromagnetic induction).

2. When a magnet is plunged in and out of a coil of wire, voltage is induced in the coil. If the rate of the in-and-out motion of the magnet is doubled, the induced voltage (doubles) (halves) (remains the same).

   If instead the number of loops in the coil is doubled, the induced voltage (doubles) (halves) (remains the same).

3. A rapidly changing magnetic field in any region of space induces a rapidly changing (electric field) (magnetic field) (gravitational field)

   which in turn induces a rapidly changing (magnetic field) (electric field) (baseball field).

   This generation and regeneration of electric and magnetic fields make up (electromagnetic waves) (sound waves) (both of these).

11. The illustration below is similar to Figure 12.2 in your textbook. Iron filings trace out patterns of magnetic field lines about a bar magnet. In the field are some magnetic compasses. The compass needle in only one compass is shown. Draw in the needles with proper orientation in the other compasses.

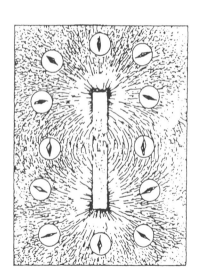

12. The illustration below is similar to Figure 12.12 (center) in your textbook. Iron filings trace out the magnetic field pattern about the loop of current-carrying wire. Draw in the compass needle orientations for all the compasses.

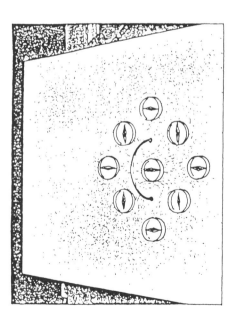

CONCEPTUAL PHYSICAL SCIENCE EXPLORATIONS
Chapter 13 Waves and Sound
Vibrations and Waves

1. A sine curve that represents a transverse wave is drawn below. With a ruler, measure the wavelength and amplitude of the wave.

a. Wavelength = __6.7 cm__

b. Amplitude = __1.5 cm__

2. A kid on a playground swing makes a complete to-and-fro swing each 2 seconds. The frequency of swing is

(0.5 hertz) (1 hertz) (2 hertz)

and the period is

(0.5 second) (1 second) (2 seconds)

3. *Complete the statements.*

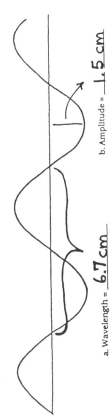

A MARINE WEATHER STATION REPORTS WAVES ALONG THE SHORE THAT ARE 8 SECONDS APART. THE FREQUENCY OF THE WAVES IS THEREFORE __1/8__ HERTZ.

THE PERIOD OF A 440-HERTZ SOUND WAVE IS __1/440__ SECOND.

4. The annoying sound from a mosquito is produced when it beats its wings at the average rate of 600 wingbeats per second.

a. What is the frequency of the sound waves? __600 Hz__

b. What is the wavelength? (Assume the speed of sound is 340 m/s.) __0.57 m__

$$\lambda = \frac{340 \text{ m/s}}{600 \text{ Hz}}$$

Vibrations and Waves continued:

5. A rapid-fire pellet gun fires 10 rounds per second. The speed of the pellets is 300 m/s. __30 m__

a. What is the distance in the air between the flying pellets? __30 m__

b. What happens to the distance between the pellets if the rate of fire is increased? __DISTANCE BETWEEN BULLETS DECREASES__

6. Consider a wave generator that produces 10 pulses per second. The speed of the waves is 300 cm/s.

a. What is the wavelength of the waves? __30 cm__

b. What happens to the wavelength if the frequency of pulses is increased? __λ DECREASES, JUST AS DISTANCE BETWEEN BULLETS IN #5 DECREASES__

7. The bird at the right watches the waves. If the portion of a wave between 2 crests passes the pole each second, what is the speed of the wave?

$v = f\lambda = 2 \times 1 \text{ m} = 2 \text{ m/s}$

What is its period? $T = \frac{1}{f} = \frac{1}{2} = 0.5 \text{ s}$

8. If the distance between crests in the above question were 1.5 meters apart, and 2 crests pass the pole each second, what would be the speed of the wave?

$v = f\lambda = 2 \times 1.5 \text{ m} = 3 \text{ m/s}$

What would be its period? __SAME (0.5 s)__

9. When an automobile moves toward a listener, the sound of its horn seems relatively

(low pitched) (normal) (**high pitched**)

and when moving away from the listener, its horn seems

(**low pitched**) (normal) (high pitched)

10. The changed pitch of the Doppler effect is due to changes in

(wave speed) (**wave frequency**)

CONCEPTUAL PHYSICAL SCIENCE EXPLORATIONS

Chapter 13 Waves and Sound
Sound

Sound is the only thing we hear!

1. Two major classes of waves are *longitudinal* and *transverse*. Sound waves are (longitudinal) (transverse)

2. The frequency of a sound signal refers to how frequently the vibrations occur. A high-frequency sound is heard at a high (pitch) (wavelength) (speed)

3. The sketch below shows a snapshot of the compressions and rarefactions of the air in a tube as the sound moves toward the right. The dots represent molecules. With a ruler the wavelength of the sound wave is measured to be __2.5__ cm.

4. Compared to the wavelengths of high-pitched sounds, the wavelengths of low-pitched sounds are (long) (short)

5. Suppose you set your watch by the sound of the noon whistle from a factory 3 km away.

LET'S SEE, FROM $v = \frac{d}{t}$ ......
$t = \frac{d}{v} = \frac{3000 \text{ m}}{340 \text{ m/s}}$

PHYSICS

a. Compared to the correct time, your watch will be (behind) (ahead)

b. It will differ from the correct time by (3 seconds) (6 seconds) (9 seconds)

Sound continued:

6. Sound waves travel fastest in (solids) (liquids) (gases) (...same speed in each)

7. If the child's natural frequency of swinging is once each 4 seconds, for maximum amplitude the man should push at a rate of once each (2 seconds) (4 seconds) (8 seconds)

8. If the man in Question 7 pushes in the same direction twice as often, his pushes (will) (will not) be effective because (the swing will be pushed twice as often in the right direction) (every other push will oppose the motion of the swing)

9. The frequency of the tuning fork is 440 hertz. It will NOT be forced into vibration by a sound of (220 hertz) (440 hertz) (880 hertz)

10. Beats are the result of the alternate cancellation and reinforcement of two sound waves of (the same frequency) (slightly different frequencies)

11. Beat frequency equals the difference between frequencies. So if two notes with frequencies of 66 and 70 Hz are sounded together, the resulting beat frequency is (4 hertz) (68 hertz) (136 hertz)

12. The accepted value for the speed of sound in air is 332 m/s at 0°C. The speed of sound in air increases 0.6 m/s for each Celsius degree above zero. Compute the speed of sound at the temperature of the room you are now in.

    SUPPOSE ROOM TEMP IS 22°C. THEN 22×0.6 m/s $= 13.2$ m/s.
    SO AT 22°C, THE SPEED OF SOUND IS ABOUT $332 + 13 = 345$ m/s.

CONCEPTUAL PHYSICAL SCIENCE **EXPLORATIONS**

Chapter 13 Waves and Sound

Shock Waves

The cone-shaped shock wave produced by a supersonic aircraft is actually the result of overlapping spherical waves of sound, as indicated in Figures 13.31 and 13.32 in your textbook. Sketches *a*, *b*, *c*, *d*, and *e*, at the left show the "animated" growth of only one of the many spherical sound waves (shown as an expanding circle in the two-dimensional sketch). The circle originates when the aircraft is in the position shown in *a*. Sketch *b* shows both the growth of the circle and position of the aircraft at a later time. Still later times are shown in *c*, *d*, and *e*. Note that the circle grows and the aircraft moves farther to the right. Note also that the aircraft is moving farther than the sound wave. This is because the aircraft is moving faster than sound.

Careful examination will reveal how fast the aircraft is moving compared to the speed of sound. Sketch *e* shows that in the same time the sound travels from O to A, the aircraft has traveled from O to B — twice as far. You can check this with a ruler.

Circle the answer.

1. Inspect sketches *b* and *d*. Has the aircraft traveled twice as far as sound in the same time in these positions also?

   (yes)   (no)

2. For greater speeds, the angle of the shock wave would be

   (wider)   (the same)   (narrower)

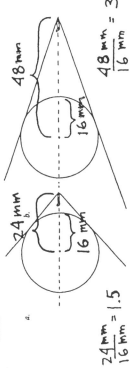

DURING THE TIME THAT SOUND TRAVELS FROM O TO A, THE PLANE TRAVELS TWICE AS FAR ... FROM O TO B.

SO IT'S FLYING AT TWICE THE SPEED OF SOUND!

Shock Waves continued:

3. Use a ruler to estimate the speeds of the aircraft that produce the shock waves in the two sketches below.

$$\frac{24\ mm}{16\ mm} = 1.5$$

$$\frac{48\ mm}{16\ mm} = 3.0$$

Aircraft *a* is traveling about __1.5__ times the speed of sound.

Aircraft *b* is traveling about __3.0__ times the speed of sound.

4. Draw your own circle (anywhere) and estimate the speed of the aircraft to produce the shock wave shown below. It will be helpful to use a coin to draw your circle. Then move it and count the number of diameters that make up your shock wave.

   ANY CIRCLE WILL DO. HERE WE'VE USED A QUARTER AND FOUND 2 ADDITIONAL ONES REACH APEX.

   FOR ANY CIRCLE, THE DISTANCE TO THE APEX WILL BE 5 TIMES GREATER THAN RADIUS OF THE CIRCLE.

   The speed is about __5__ times the speed of sound.

5. In the space below, draw the shock wave made by a supersonic missile that travels at four times the speed of sound.

CONCEPTUAL PHYSICAL SCIENCE EXPLORATIONS

Chapter 14 Light and Color
Light

LIGHT IS THE ONLY THING WE SEE!

1. Study Figure 14.3 in your textbook and answer the following:

a. Which has the longer *wavelengths*, radio waves or light waves?

   RADIO WAVES

b. Which has the longer *wavelengths*, light waves or gamma rays?

   LIGHT WAVES

c. Which has the higher *frequencies*, ultraviolet or infrared waves?

   ULTRAVIOLET WAVES

d. Which has the higher *frequencies*, ultraviolet waves or gamma rays?

   GAMMA RAYS

2. Carefully study Section 14.2 in your textbook and answer the following:

a. Exactly what do vibrating electrons emit?

   ENERGY THAT IS CARRIED IN AN ELECTROMAGNETIC WAVE.

b. When ultraviolet light shines on glass, what does it do to electrons in the glass structure?

   ULTRAVIOLET LIGHT CAUSES ELECTRONS TO VIBRATE IN RESONANCE WITH THE ULTRAVIOLET LIGHT.

c. When energetic electrons in the glass structure vibrate against neighboring atoms, what happens to the energy of vibration?

   THE ENERGY OF VIBRATION BECOMES HEAT.

d. What happens to the energy of a vibrating electron that does not collide with neighboring atoms?

   THE ENERGY IS EMITTED AS LIGHT.

e. Which range of light frequencies, visible or ultraviolet, are absorbed in glass?

   ULTRAVIOLET

f. Which range of light frequencies, visible or ultraviolet, are transmitted through glass?

   VISIBLE

g. How is the speed of light in glass affected by the succession of time delays that accompany the absorption and re-emission of light from atom to atom in the glass?

   TIME DELAYS LOWER AVERAGE SPEED

h. How does the speed of light compare in water, glass, and diamond?

   IN ORDER OF SLOWNESS: DIAMOND, GLASS, WATER

CONCEPTUAL PHYSICAL SCIENCE EXPLORATIONS

Chapter 14 Light and Color
Color

The sketch to the right shows the shadow of a teacher in front of a white screen in a darkened room. The light source is red, so the screen looks red and the shadow looks black. Color the sketch with colored markers, or label the colors with pen or pencil.

RED   BLACK

A green lamp is added and makes a second shadow. The formerly black shadow cast by the red light is no longer black, but is illuminated with green light. So it is green. Color or mark it green. The shadow cast by the green lamp is not black, because it is illuminated with the red light. Color or mark its color. The background receives a mixture of red and green light. Figure out what color the background will appear, then color or label it.

RED   YELLOW   GREEN   RED   GREEN

A blue lamp is added and three shadows appear. Color or label the appropriate colors of the shadows and the background.

MAGENTA YELLOW WHITE CYAN RED BLUE GREEN

The lamps are placed closer together so the shadows overlap. Indicate the colors of all screen areas.

MAGENTA RED GREEN YELLOW WHITE CYAN RED BLUE GREEN

CONCEPTUAL PHYSICAL SCIENCE EXPLORATIONS

Chapter 15 Reflection and Refraction

Reflection

1. Light from a flashlight shines on a mirror and illuminates one of the cards. Draw the reflected beam to indicate the illuminated card.

2. A periscope has a pair of mirrors in it. Draw the light path from the object "O" to the eye of the observer.

3. The ray diagram below shows the extension of one of the reflected rays from the plane mirror. Complete the diagram by (1) carefully drawing the three other reflected rays, and (2) extending them behind the mirror to locate the image of the flame. (Assume the candle and image are viewed by an observer on the left.)

Color continued:

If you have colored markers, have a go at these.

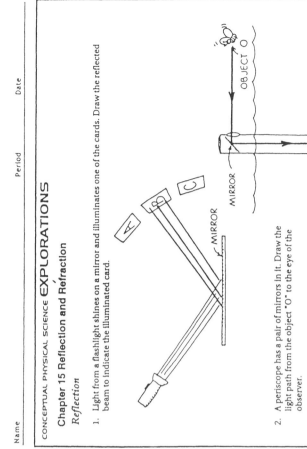

CONCEPTUAL PHYSICAL SCIENCE **EXPLORATIONS**

Chapter 15 Reflection and Refraction

Refraction

1. A pair of toy cart wheels are rolled obliquely from a smooth surface onto two plots of grass — a rectangular plot as shown at the left, and a triangular plot as shown at the right. The ground is on a slight incline so that after slowing down in the grass, the wheels speed up again when emerging on the smooth surface. Finish each sketch and show some positions of the wheels inside the plots and on the other side. Clearly indicate their paths and directions of travel.

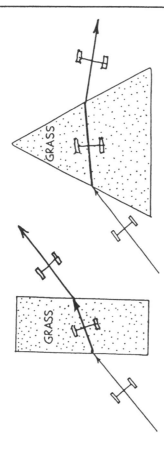

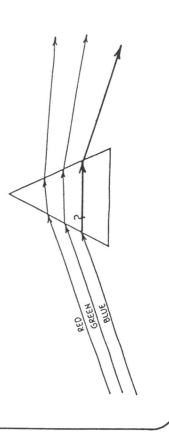

2. Red, green, and blue rays of light are incident upon a glass prism as shown. The average speed of red light in the glass is less than in air, so the red ray is refracted. When it emerges into the air it regains its original speed and travels in the direction shown. Green light takes longer to get through the glass. Because of its slower speed it is refracted as shown. Blue light travels even slower in glass. Complete the diagram by estimating the path of the blue ray.

Reflection continued:

4. The ray diagram below shows the reflection of one of the rays that strikes the parabolic mirror. Notice that the law of reflection is obeyed, and the angle of incidence (from the normal, the dashed line) equals the angle of reflection (from the normal). Complete the diagram by drawing the reflected rays of the other three rays that are shown. (Do you see why parabolic mirrors are used in automobile headlights?)

5. A girl takes a photograph of the bridge as shown. Which of the two sketches correctly shows the reflected view of the bridge? Defend your answer.

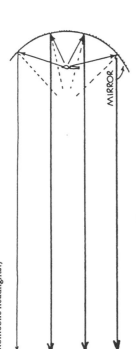

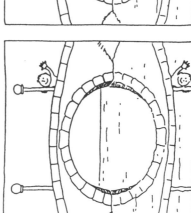

The right-side view is correct, showing the underside of bridge, or what your eye would see if as far below the reflecting surface as it is above. (Place a mirror on the floor in front of a table. Students will see that the reflected view of the table shows its bottom.)

CONCEPTUAL PHYSICAL SCIENCE EXPLORATIONS

Chapter 15 Reflection and Refraction
Measuring the Diameter of the Sun with a Ruler

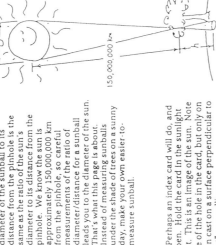

150,000,000 km

Look carefully at the round spots of light on the shady ground beneath trees. These are *sunballs*, which are images of the sun. They are cast by openings between leaves in the trees that act as pinholes. (Did you make a pinhole "camera" back in middle school?) Large sunballs, several centimeters in diameter or so, are cast by openings that are relatively high above the ground, while small ones are produced by closer "pinholes." The interesting point is that the ratio of the diameter of the sunball to its distance from the pinhole is the same as the ratio of the sun's diameter to its distance from the pinhole. We know the sun is approximately 150,000,000 km from the pinhole, so careful measurements of the ratio of diameter/distance for a sunball leads you to the diameter of the sun. That's what this page is about. Instead of measuring sunballs under the shade of trees on a sunny day, make your own easier-to-measure sunball.

1. Poke a small hole in a piece of card. Perhaps an index card will do, and poke the hole with a sharp pencil or pen. Hold the card in the sunlight and note the circular image that is cast. This is an image of the sun. Note that its size doesn't depend on the size of the hole in the card, but only on its distance. The image is a circle when cast on a surface perpendicular to the rays — otherwise it's "stretched out" as an ellipse.

2. Try holes of various shapes, say a square hole, or a triangular hole. What is the shape of the image when its distance from the card is large compared with the size of the hole? Does the shape of the pinhole make a difference?

CIRCLE

NO, THE IMAGE REMAINS A CIRCLE (ONLY BRIGHTNESS VARIES).

3. Measure the diameter of a small coin. Then place the coin on a viewing area that is perpendicular to the sun's rays. Position the card so the image of the sunball exactly covers the coin. Carefully measure the distance between the coin and the small hole in the card. Complete the following:

$$\frac{\text{Diameter of sunball}}{\text{Distance to pinhole}} = \frac{1}{108}$$

With this ratio, estimate the diameter of the sun. Show your work on a separate piece of paper. $\dfrac{D_{sun}}{150,000,000\ km} = \dfrac{1}{108}$; $D_{sun} = \dfrac{1}{108}(150,000,000\ km) = 1,900,000\ km$

4. If you did this on a day when the sun is partially eclipsed, what shape of image would you expect to see?

UPSIDE-DOWN CRESCENTS... IMAGE OF A PARTIALLY-ECLIPSED SUN

WHAT SHAPE DO SUNBALLS HAVE DURING A PARTIAL ECLIPSE OF THE SUN?

Refraction continued:

3. The sketch to the right shows a light ray moving from air into water, at 45° to the normal. Which of the three rays indicated with capital letters is most likely the light ray that continues inside the water?

C

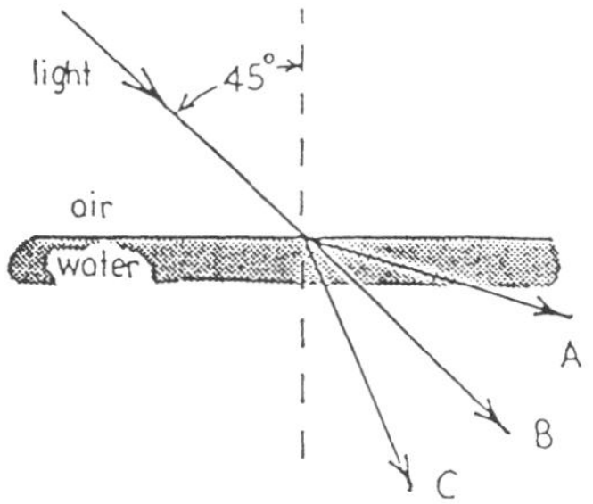

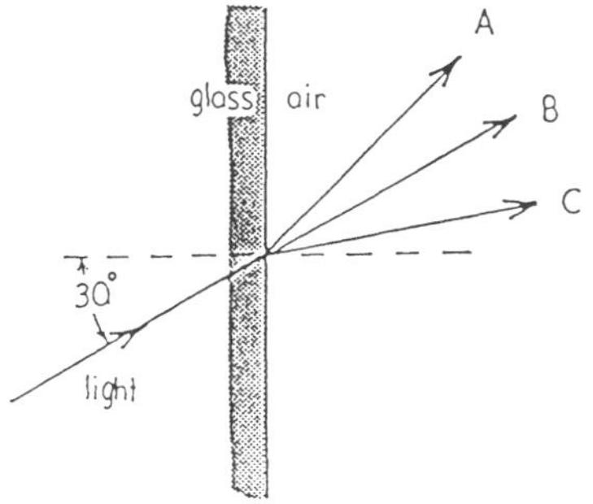

4. The sketch on the left shows a light ray moving from glass into air, at 30° to the normal. Which of the three is most likely the light ray that continues in the air?

A

5. To the right, at 40° to the normal. Which of the three rays is most likely the light ray that travels in the air after emerging from the opposite side of the block?

A

Sketch the path the light would take inside the glass.

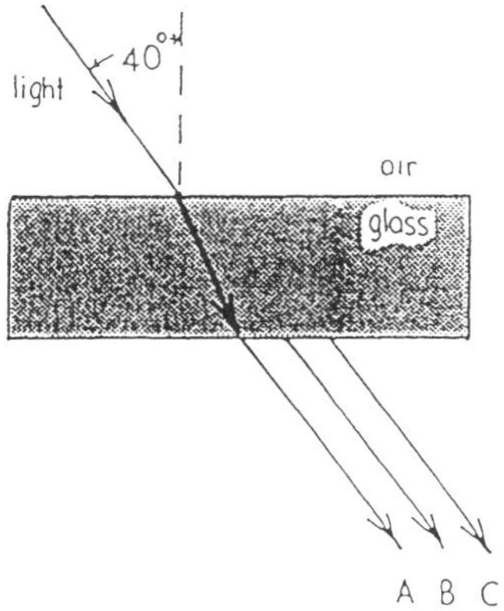

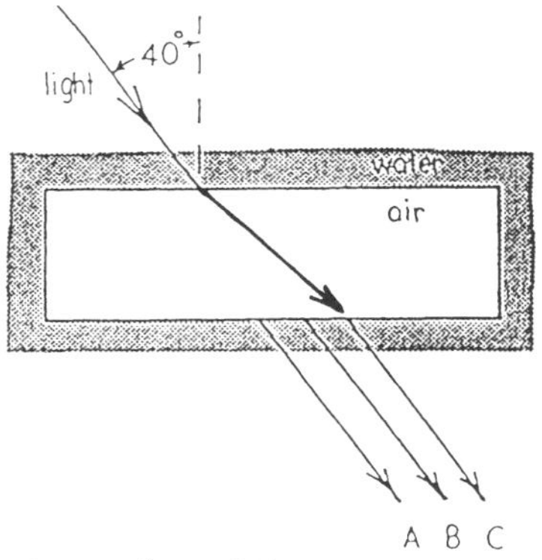

6. To the left, a light ray is shown moving from water into a rectangular block of air (inside a thin-walled plastic box), at 40° to the normal. Which of the three rays is most likely the light ray that continues into the water on the opposite side of the block?

C

Sketch the path the light would take inside the air.

179

CONCEPTUAL PHYSICAL SCIENCE EXPLORATIONS
Chapter 16 Properties of Light
Interference

1. Sketch an identical wave upon the one below, that is in phase. Then sketch the resulting wave.

2. Here is the wave again. This time sketch an identical wave upon it that is out of phase. Sketch the result.

3. Figure 16.9 from your text is repeated below. Carefully count the number of wavelengths (same as the number of wave crests) along the following paths between the slits and the screen.

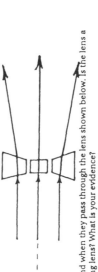

a LIGHT

b DARK

c LIGHT

DARK

LIGHT

a. Number of wavelengths between slit A and point a = **10.5**

b. Number of wavelengths between slit B and point a = **11.5**

c. Number of wavelengths between slit A and point b = **10.0**

d. Number of wavelengths between slit B and point b = **10.5**

e. Number of wavelengths between slit A and point c = **10.0**

f. Number of wave crests between slit B and point c = **10.0**

When the number of wavelengths along each path is the same or differs by one or more whole wavelengths, interference is (constructive) (destructive)

and when the number of wavelengths differ by a half wavelength (or odd multiples of a half wavelength), interference is (constructive) (destructive).

65

Chapter 15 Reflection and Refraction
Lenses

Rays of light bend as shown when passing through the glass blocks.

1. Show how light rays bend when they pass through the arrangement of glass blocks shown below.

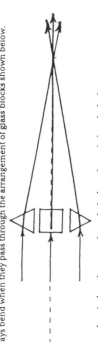

2. Show how light rays bend when they pass through the lens shown below. Is the lens a converging or a diverging lens? What is your evidence?
CONVERGING, AS EVIDENT IN THE CONVERGING RAYS.

3. Show how light rays bend when they pass through the arrangement of glass blocks shown below.

4. Show how light rays bend when they pass through the lens shown below. Is the lens a converging or a diverging lens? What is your evidence?
DIVERGING LENS, AS EVIDENT IN THE DIVERGING RAYS.

Being able to see is wonderful—not to be taken lightly!

64

CONCEPTUAL PHYSICAL SCIENCE EXPLORATIONS

Chapter 16 Properties of Light
Polarization

The amplitude of a light wave has magnitude and direction, and can be represented by a **vector**. Polarized light vibrates in a single direction and is represented by a single vector. To the left the single vector represents vertically polarized light. The vibrations of nonpolarized light are equal in all directions. There are as many vertical components as horizontal components. The pair of perpendicular vectors to the right represents nonpolarized light.

1. In the sketch below nonpolarized light from a flashlight strikes a pair of Polaroid filters.

NON-POLARIZED LIGHT VIBRATES IN ALL DIRECTIONS

HORIZONTAL AND VERTICAL COMPONENTS

VERTICAL COMPONENT PASSES THROUGH FIRST POLARIZER

...AND THE SECOND

VERTICAL COMPONENT DOES NOT PASS THROUGH THIS SECOND POLARIZER

a. Light is transmitted by a pair of Polaroids when their axes are

(aligned)   (crossed at right angles)

and light is blocked when their axes are

(aligned)   (crossed at right angles)

b. Transmitted light is polarized in a direction

(the same as)   (different than)   the polarization axis of the filter.

2. Consider the transmission of light through a pair of Polaroids with polarization axes at 45° to each other. Although in practice the Polaroids are one atop the other, we show them spread out side by side below. From left to right: (a) Nonpolarized light is represented by its horizontal and vertical components. (b) These components strike filter A. (c) The vertical component is transmitted, and (d) falls upon filter B. This vertical component is not aligned with the polarization axis of filter B, but it has a component that is — component t; (e) which is transmitted.

(a)   (b)   (c)   (d)   (e)

A   B

45°

a. The amount of light that gets through Filter B, compared to the amount that gets through Filter A is

(more)   (less)   (the same)

b. The component perpendicular to t that falls on Filter B is

(also transmitted)   (absorbed)

Chapter 16 Properties of Light
Wave-Particle Duality

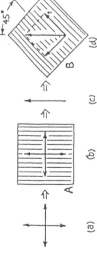

E=hf

1. To say that light is quantized means that light is made up of

(elemental units)

(waves)

2. Compared to photons of low-frequency light, photons of higher-frequency light have more

(energy)

(speed)

(quanta)

3. The photoelectric effect supports the

(wave model of light)

(particle model of light)

4. The photoelectric effect is evident when light shone on certain photosensitive materials ejects

(photons)

(electrons)

5. The photoelectric effect is more effective with violet light than with red light because the photons of violet light

(resonate with the atoms in the material)

(deliver more energy to the material)

(are more numerous)

6. According to the wave-particle model of light, light travels as a

(wave and hits like a particle)

(particle and hits like a wave)

7. According to the wave-particle model of matter, a beam of light and a beam of electrons

(are fundamentally different)   (are similar)

8. Whereas in the everyday macroworld the study of motion is called mechanics, in the microworld the study of quanta is called

(quantum mechanics)

(Newtonian mechanics)

181

CONCEPTUAL PHYSICAL SCIENCE EXPLORATIONS
Chapter 17 Atoms and the Periodic Table
Subatomic Particles

Three fundamental particles of the atom are the __PROTON__, __NEUTRON__ and __ELECTRON__.

At the center of each atom lies the atomic __NUCLEUS__, which consists of __PROTONS__, and

__NEUTRONS__. The atomic number refers to the number of __PROTONS__ in the nucleus. All atoms of the same element have the same number of __PROTONS__, hence, the same atomic number.

Isotopes are atoms that have the same number of __PROTONS__ but a different number of

__NEUTRONS__. An isotope is identified by its atomic mass number, which is the total number of __PROTONS__ and __NEUTRONS__ the nucleus. A carbon isotope that has 6 __PROTONS__ and 6 __NEUTRONS__ is identified as carbon-12, where 12 is the atomic mass number. A carbon isotope

having 6 __PROTONS__ and 8 __NEUTRONS__ on the other hand, is carbon-14.

1. Complete the following table:

Isotope	Number of... Electrons	Protons	Neutrons
Hydrogen-1	1	1	0
Chlorine-36	17	17	19
Nitrogen-14	7	7	7
Potassium-40	19	19	21
Arsenic-75	33	33	42
Gold-197	79	79	118

2. Which results in a more valuable product — *adding* or *subtracting* protons from gold nuclei?
SUBTRACT FOR PLATINUM (MORE VALUABLE)

3. Which has more mass, a helium atom or a neon atom?
NEON

4. Which has a greater number of atoms, a gram of helium or a gram of neon?
HELIUM!

I LIKE THE WAY YOUR ATOMS ARE PUT TOGETHER!
:SIGH:

CONCEPTUAL PHYSICAL SCIENCE EXPLORATIONS
Chapter 17 Atoms and the Periodic Table
Melting Points of the Elements

There is a remarkable degree of organization in the periodic table. As discussed in your textbook, elements within the same atomic group (vertical column) share similar properties. Also, the chemical reactivity of an element can be deduced from its position in the periodic table. Two additional examples of the periodic table's organization are the melting points and densities of the elements.

The periodic table below shows the melting points of nearly all the elements. Note the melting points are not randomly oriented, but, with only a few exceptions, either gradually increase or decrease as you move in any particular direction. This can be clearly illustrated by color coding each element according to its melting point.

Use colored pencils to color in each element according to its melting point. Use the suggested color legend. Color lightly so that symbols and numbers are still visible.

Color	Temperature Range, °C	Color	Temperature Range, °C
Violet	-273 — -50	Yellow	1400 — 1900
Blue	-50 — 300	Orange	1900 — 2900
Cyan	300 — 700	Red	2900 — 3500
Green	700 — 1400		

Melting Points of the Elements (°C)

1	2	3	4	5	6	7	8	9	10	11	12	13	14	15	16	17	18
H -259																	He -272
Li 180	Be 1278											B 2079	C 3550	N -210	O -218	F -219	Ne -248
Na 97	Mg 648											Al 660	Si 1410	P 44	S 113	Cl -100	Ar -189
K 63	Ca 839	Sc 1541	Ti 1660	V 1890	Cr 1857	Mn 1244	Fe 1535	Co 1495	Ni 1453	Cu 1083	Zn 419	Ga 30	Ge 937	As 817	Se 217	Br -7	Kr -156
Rb 39	Sr 769	Y 1522	Zr 1852	Nb 2468	Mo 2617	Tc 2172	Ru 2310	Rh 1966	Pd 1554	Ag 961	Cd 320	In 156	Sn 231	Sb 630	Te 449	I 113	Xe -111
Cs 28	Ba 725	Lo 921	Hf 2227	Ta 2996	W 3410	Re 3180	Os 3045	Ir 2440	Pt 1772	Au 1064	Hg -38	Tl 303	Pb 327	Bi 271	Po 254	At 302	Rn -71
Fr 27	Ra 700	Ac 1050															

— TUNGSTEN (W 3410)

Lanthanides	Ce 799	Pr 931	Nd 1021	Pm 1168	Sm 1077	Eu 822	Gd 1313	Tb 1356	Dy 1412	Ho 1474	Er 1159	Tm 1545	Yb 819	Lu 1663
Actinides	Th 1750	Pa 1600	U 1132	Np 640	Pu 641	Am 994	Cm 1340	Bk --	Cf --	Es --	Fm --	Md --	No --	Lr --

1. Which elements have the highest melting points?
THE ONES CLOSER TO TUNGSTEN

2. Which elements have the lowest melting points?
ELEMENTS TOWARD UPPER RIGHT

3. Which atomic groups tend to go from higher to lower melting points reading from top to bottom? (Identify each group by its group number).
1, 2, 3, 12, 13, 14

4. Which atomic groups tend to go from lower to higher melting points reading from top to bottom?
4 THROUGH 10 AND 15 THROUGH 18

CONCEPTUAL PHYSICAL SCIENCE EXPLORATIONS

Chapter 18 Atomic Models
Losing Valence Electrons

The shell model described in Section 18.4 can be used to explain a wide variety of properties of atoms. Using the shell model, for example, we can explain how atoms within the same group tend to lose (or gain) the same number of electrons. Let's consider the case of three group 1 elements: lithium, sodium, and potassium. Look to a periodic table and find the nuclear charge of each of these atoms:

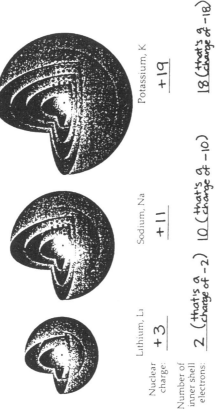

Lithium, Li	Sodium, Na	Potassium, K
Nuclear charge: $+3$	$+11$	$+19$
Number of inner shell electrons: 2 (that's a charge of -2)	10 (that's a charge of -10)	18 (that's a charge of -18)

How strongly the valence electron is held to the nucleus depends on the strength of the nuclear charge—the stronger the charge, the stronger the valence electron is held. There's more to it, however, because inner-shell electrons weaken the attraction outer-shell electrons have for the nucleus. The valence shell in lithium, for example, doesn't experience the full effect of three protons. Instead, it experiences a diminished nuclear charge of about +1. We get this by subtracting the number of inner-shell electrons from the actual nuclear charge. What do the valence electrons for sodium and potassium experience?

Diminished nuclear charge:	$(+3 - 2 = +1)$ about $+1$	$(+11 - 10 = +1)$ about $+1$	$(+19 - 18 = +1)$ about $+1$

Question: Potassium has a nuclear charge many times greater than that of lithium. Why is it actually *easier* for a potassium atom to lose its valence electron than it is for a lithium atom to lose its valence electron?

Potassium's valence electron is much farther from the nucleus. Because the electric force decreases with distance, the +1 charge for potassium's valence electron is not so effective at holding to the atom. Hence, it is easily lost.

Hint: Remember from Chapter 11 what happens to the electric force as distance is increased!

Chapter 17 Atoms and the Periodic Table
Densities of the Elements

The periodic table below shows the densities of nearly all the elements. As with the melting points, the densities of the elements either gradually increase or decrease as you move in any particular direction. Use colored pencils to color in each element according to its density. Shown below is a suggested color legend. Color lightly so that symbols and numbers are still visible. (Note: All gaseous elements are marked with an asterisk and should be the same color. Their densities, which are given in units of g/L, are much less than the densities non-gaseous elements, which are given in units of g/mL.)

Color	Density (g/mL)	Color	Density (g/mL)
Violet	gaseous elements	Yellow	16 — 12
Blue	5 — 0	Orange	20 — 16
Cyan	8 — 5	Red.	23 — 20
Green	12 — 8		

Densities of the Elements
(g/mL)

1	2	3	4	5	6	7	8	9	10	11	12	13	14	15	16	17	18
H* 0.09																	He* 0.18
Li 0.5	Be 1.8											B 2.3	C 2.0	N* 1.25	O* 1.43	F* 1.70	Ne* 0.90
Na 1.0	Mg 1.7											Al 2.7	Si 2.3	P 1.8	S 2.1	Cl* 3.21	Ar* 1.78
K 0.9	Ca 1.6	Sc 3.0	Ti 4.5	V 6.1	Cr 7.2	Mn 7.3	Fe 7.8	Co 8.9	Ni 8.9	Cu 9.0	Zn 7.1	Ga 6.1	Ge 5.3	As 5.7	Se 4.8	Br* 7.59	Kr* 3.73
Rb 1.5	Sr 2.5	Y 4.5	Zr 6.5	Nb 8.5	Mo 6.8	Tc 11.5	Ru 12.4	Rh 12.4	Pd 12.0	Ag 10.5	Cd 8.7	In 7.3	Sn 5.7	Sb 6.7	Te 6.2	I* 4.9	Xe* 5.89
Cs 1.9	Ba 3.5	La 6.2	Hf 13.3	Ta 16.6	W 19.3	Re 21.0	Os (22.6)	Ir 22.4	Pt 21.5	Au 18.9	Hg 13.5	Tl 11.9	Pb 11.4	Bi 9.7	Po 9.3	At* --	Rn* 9.73
Fr --	Ra --	Ac 10.1	Unq --	Unp --	Unh --	Uns --	Uno --	Une --									

OSMIUM

* density of gaseous phase in g/L

Lanthanides:	Ce 6.7	Pr 6.7	Nd 6.8	Pm 7.2	Sm 7.5	Eu 5.2	Gd 7.9	Tb 8.2	Dy 8.6	Ho 8.8	Er 9.1	Tm 9.3	Yb 6.9	Lu 9.8
Actinides:	Th 11.7	Pa 15.4	U 19.0	Np 20.1	Pu 19.8	Am 13.7	Cm 13.5	Bk 14	Cf --	Es --	Fm --	Md --	No --	Lr --

1. Which elements are the most dense?

    THE ONES CLOSER TO OSMIUM, OS

2. How variable are the densities of the lanthanides compared to the densities of the actinides?

    THE ACTINIDES ARE MUCH MORE VARIABLE

3. Which atomic groups tend to go from higher to lower densities reading from top to bottom? (Identify each group by its group number).

    NONE

4. Which atomic groups tend to go from lower to higher densities reading from top to bottom?

    ALL

CONCEPTUAL PHYSICAL SCIENCE EXPLORATIONS

Chapter 19 Radioactivity
The Atomic Nucleus and Radioactivity

1. *Complete the following statements.*

a. A lone neutron spontaneously decays into a proton plus an ___ELECTRON___.

b. Alpha and beta rays are made of streams of particles, whereas gamma rays are streams of ___PHOTONS___.

c. An electrically charged atom is called an ___ION___.

d. Different ___ISOTOPES___ of an element are chemically identical but differ in the number of neutrons in the nucleus.

e. Transuranic elements are those beyond atomic number ___92___.

f. If the amount of a certain radioactive sample decreases by half in four weeks, in four more weeks the amount remaining should be ___ONE-FOURTH___ the original amount.

g. Water from a natural hot spring is warmed by ___RADIOACTIVITY___ inside the Earth.

2. The gas in the little girl's balloon is made up of former alpha and beta particles produced by radioactive decay.

a. If the mixture is electrically neutral, how many more beta particles than alpha particles are in the balloon?
___THERE ARE TWICE AS MANY BETA PARTICLES AS ALPHA PARTICLES___

b. Why is your answer not "same"?
___AN ALPHA HAS DOUBLE CHARGE; THE CHARGE OF 2 BETAS = MAGNITUDE OF CHARGE OF 1 ALPHA PARTICLE.___

c. Why are the alpha and beta particles no longer harmful to the child?
___THEY HAVE LONG LOST THEIR HIGH KE, WHICH IS NOW REDUCED TO THE THERMAL ENERGY OF RANDOM MOLECULAR MOTION.___

d. What element does this mixture make?
___HELIUM___

Chapter 19 Radioactivity
Natural Transmutation

Fill in the decay-scheme diagram below, similar to that shown in Figure 19.11 in the textbook but beginning with U-235 and ending up with an isotope of lead. Use the table at the left, and identify each element in the series with its chemical symbol.

Step	Particle emitted
1	Alpha
2	Beta
3	Alpha
4	Alpha
5	Beta
6	Alpha
7	Alpha
8	Alpha
9	Beta
10	Alpha
11	Beta
12	Stable

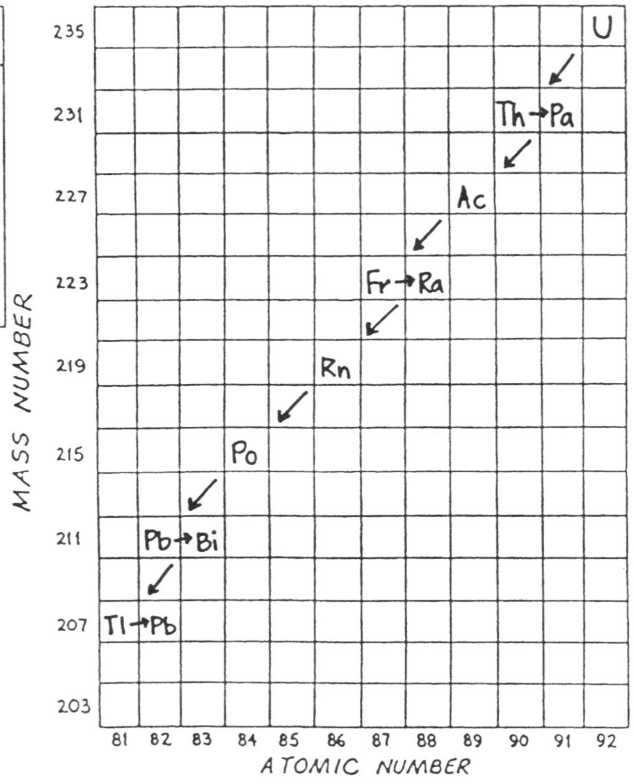

What isotope is the final product? ___$^{207}_{82}Pb$ (LEAD-207)___

CONCEPTUAL PHYSICAL SCIENCE **EXPLORATIONS**

Chapter 20 Nuclear Fission and Fusion
Nuclear Fission and Fusion

1. Complete the table for a chain reaction in which two neutrons from each step individually cause a new reaction.

EVENT	1	2	3	4	5	6	7
NO. OF REACTIONS	1	2	4	8	16	32	64

2. Complete the table for a chain reaction where three neutrons from each reaction cause a new reaction.

EVENT	1	2	3	4	5	6	7
NO. OF REACTIONS	1	3	9	27	81	243	729

3. Complete these beta reactions, which occur in a breeder reactor.

$$^{239}_{92}U \rightarrow \underline{^{239}_{93}Np} + ^{0}_{-1}e$$

$$^{239}_{93}Np \rightarrow \underline{^{239}_{94}Pu} + ^{0}_{-1}e$$

4. Complete the following fission reactions.

$$^{1}_{0}n + ^{235}_{92}U \rightarrow ^{143}_{54}Xe + ^{90}_{38}Sr + \underline{3}\ (^{1}_{0}n)$$

$$^{1}_{0}n + ^{235}_{92}U \rightarrow ^{152}_{60}Nd + \underline{^{80}_{32}Ge} + 4\ (^{1}_{0}n)$$

$$^{1}_{0}n + ^{239}_{94}Pu \rightarrow ^{141}_{\underline{54}}Xe + ^{97}_{40}Zr + 2\ (^{1}_{0}n)$$

5. Complete the following fusion reactions.

$$^{2}_{1}H + ^{2}_{1}H \rightarrow ^{3}_{2}He + \underline{^{1}_{0}n}$$

$$^{2}_{1}H + ^{3}_{1}H \rightarrow ^{4}_{2}He + \underline{^{1}_{0}n}$$

KNOW NUKES!

CONCEPTUAL PHYSICAL SCIENCE **EXPLORATIONS**

Chapter 19 Radioactivity
Radioactive Half-Life

You and your classmates will now play the "half-life game." Each of you should have a coin to shake inside cupped hands. After it has been shaken for a few seconds, the coin is tossed on the table or on the floor. Students with tails up fall out of the game. Students who consistently show heads remain in the game. Finally everybody has tossed a tail and the game is over.

1. The graph to the left shows the decay of Radium-226 with time. Note that each 1620 years, half remains (the rest changes to other elements). In the grid below, plot the number of students left in the game after each toss. Draw a smooth curve that passes close to the points on your plot. What is the similarity of your curve with that of the curve of Radium-2226?

SHOULD BOTH DECREASE RAPIDLY

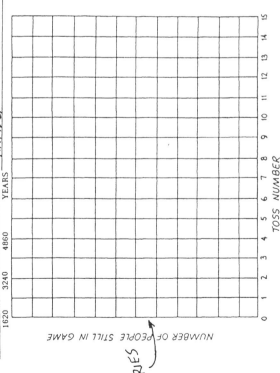

VARIES

2. Was the person to last longest in the game *lucky*, with some sort of special powers to guide the long survival? What test could you make to decide the answer to this question? TEST! REPEAT TO SEE IF "LUCKY" PERSON REMAINS LUCKY!

185

Chapter 20 Nuclear Fission and Fusion
Nuclear Reactions

Complete these nuclear reactions.

1. $^{238}_{92}U \rightarrow ^{234}_{90}Th + ^{4}_{2}\underline{He}$

2. $^{234}_{90}Th \rightarrow ^{234}_{91}Pa + ^{0}_{-1}\underline{e}$

3. $^{234}_{91}Pa \rightarrow ^{230}_{\underline{89}}\underline{Ac} + ^{4}_{2}He$

4. $^{220}_{86}Rn \rightarrow ^{216}_{\underline{84}}\underline{Po} + ^{4}_{2}He$

5. $^{216}_{84}Po \rightarrow ^{216}_{\underline{85}}\underline{At} + ^{0}_{-1}\underline{e}$

6. $^{216}_{84}Po \rightarrow ^{212}_{\underline{82}}\underline{Pb} + ^{4}_{2}He$

7. $^{210}_{83}Bi \rightarrow ^{210}_{\underline{84}}\underline{Po} + ^{0}_{-1}\underline{e}$

8. $^{1}_{0}n + ^{10}_{5}B \rightarrow ^{7}_{\underline{3}}\underline{Li} + ^{4}_{2}He$

THORIUM LATE, I OVERTHLEPT!

NUCLEAR PHYSICS.... IT'S THE SAME TO ME WITH THE FIRST TWO LETTERS INTERCHANGED!

Chapter 21 Elements of Chemistry
The Submicroscopic

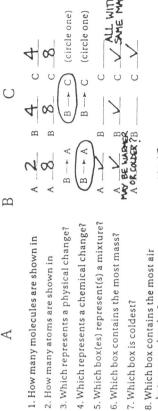

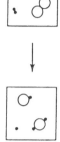

A   B   C

1. How many molecules are shown in   A $\underline{2}$ B $\underline{4}$ C $\underline{4}$

2. How many atoms are shown in   A $\underline{8}$ B $\underline{8}$ C $\underline{8}$

3. Which represents a physical change?   (B → A)   B → C   C   (circle one)

4. Which represents a chemical change?   B → A   (B → C)   C   (circle one)

5. Which box(es) represent(s) a mixture?   A ✓   B ✓   C

6. Which box contains the most mass?   A ✓   B ✓   C   ALL WITH SAME MASS ✓

7. Which box is coldest?   A MAY BE WARMER OR COLDER? B   C

8. Which box contains the most air between molecules?   A NONE   B   C   THERE IS NO AIR BETWEEN THE MOLECULES.

A   B   C

9. How many molecules are shown in   A $\underline{2}$ B $\underline{3}$ C $\underline{2}$

10. How many atoms are shown in   A $\underline{6}$ B $\underline{6}$ C $\underline{6}$

11. Which represents a physical change?   B → A   (B → C)   C

12. Which represents a chemical change?   (B → A)   B → C   C

13. Which box(es) represent(s) a mixture?   A ✓   B ✓   C

14. Which box contains the most mass?   A ✓   B   C   ALL WITH SAME MASS ✓

15. Which should take longer?   B → A   (B → C)

16. Which box most likely contains ions?   A ✓   B   C   (circle one)

NEITHER BOTH   (circle one)

ONE LESS STEP IS REQUIRED TO GO FROM B → A

CONCEPTUAL PHYSICAL SCIENCE **EXPLORATIONS**

Chapter 21 Elements of Chemistry
Balancing Chemical Equations

In a balanced chemical equation the number of times each element appears as a reactant is equal to the number of times it appears as a product. For example,

$$2\ H_2 + O_2 \longrightarrow 2\ H_2O$$

Recall that *coefficients* (the integer appearing before the chemical formula) indicate the number of times each chemical formula is to be counted and *subscripts* indicate when a particular element occurs more than once within the formula.

Check whether or not the following chemical equations are balanced.

$3\ NO \longrightarrow N_2O + NO_2$ ☑ balanced; ☐ unbalanced

$SiO_2 + 4\ HF \longrightarrow SiF_4 + 2\ H_2O$ ☑ balanced; ☐ unbalanced

$4\ NH_3 + 5\ O_2 \longrightarrow 4\ NO + 6\ H_2O$ ☑ balanced; ☐ unbalanced

Unbalanced equations are balanced by changing the coefficients. Subscripts, however, should never be changed because this changes the chemical's identity—H_2O is water, but H_2O_2 is hydrogen peroxide! The following steps may help guide you:

1. Focus on balancing only one element at a time. Start with the left-most element and modify the coefficients such that this element appears on both sides of the arrow the same number of times.

2. Move to the next element and modify the coefficients so as to balance this element. Do not worry if you incidently unbalance the previous element. You will come back to it in subsequent steps.

3. Continue from left to right balancing each element individually.

4. Repeat steps 1 - 3 until all elements are balanced.

Use the above methodology to balance the following chemical equations.

__1__ N_2O + __6__ N_2 --> __4__ O_2

__2__ $NaClO_3$ --> __2__ $NaCl$ + __3__ O_2

__3__ $MnCl_2$ + __2__ Al --> __3__ Mn + __2__ $AlCl_3$

__2__ K + __2__ H_2O --> __1__ H_2 + __2__ KOH

__2__ Al_2O_3 + __3__ C --> __4__ Al + __3__ CO_2

__4__ NH_3 + __3__ F_2 --> __3__ NH_4F + __1__ NF_3

This is just one of the many methods that chemists have developed to balance chemical equations.

Knowing how to balance a chemical equation is a useful technique, but understanding why a chemical equation needs to be balanced in the first place is far more important.

Chapter 21 Elements of Chemistry
Physical and Chemical Changes

Chemistry [sigh]

1. What distinguishes a chemical change from a physical change?

DURING A CHEMICAL CHANGE ATOM CHANGE PARTNERS

2. Based upon observations alone, why is distinguishing a chemical change from a physical change not always so straight-forward?

BOTH INVOLVE A CHANGE IN PHYSICAL APPEARANCE

Try your hand at categorizing the following processes as either chemical or physical changes. Some of these examples are debatable! Be sure to discuss your reasoning with fellow classmates or your instructor.

(circle one)

3. A cloud grows dark -------- chemical / **physical** *(circled)*
4. Leaves produce oxygen. ----- **chemical** *(circled)* / physical
5. Food coloring is added to water. ----- chemical / **physical** *(circled)*
6. Tropical coral reef dies. ----- **chemical** *(circled)* / physical
7. Dead coral reef is pounded by waves into beach sand. ----- chemical / **physical** *(circled)*
8. Oil and vinegar separate. ----- chemical / **physical** *(circled)*
9. Soda drink goes flat. ----- chemical / **physical** *(circled)*
10. Sick person develops a fever. ----- **chemical** *(circled)* / physical
11. Compost pit turns into mulch. ----- **chemical** *(circled)* / physical
12. A computer is turned on. AT THE ELECTRIC POWER PLANT → **chemical** *(circled)* / physical
13. An electrical short melts a computer's integrated circuits. ----- **chemical** *(circled)* / physical
14. A car battery runs down. ----- **chemical** *(circled)* / physical
15. A pencil is sharpened. ----- chemical / **physical** *(circled)*
16. Mascara is applied to eyelashes. ----- chemical / **physical** *(circled)*
17. Sunbather gets tan lying in the sun. ----- **chemical** *(circled)* / physical
18. Invisible ink turns visible upon heating. ----- chemical / physical
19. A light bulb burns out. ----- **chemical** *(circled)* / physical
20. Car engine consumes a tank of gasoline. ----- **chemical** *(circled)* / physical
21. B vitamins turn urine yellow. ASSUMING "XS" VITAMIN **chemical** *(circled)* / physical PASSES THROUGH BODY UNCHANGED

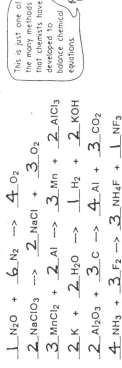

CONCEPTUAL PHYSICAL SCIENCE EXPLORATIONS

Chapter 22 Mixtures
Pure Mathematics

Using a scientist's definition of pure, identify whether each of the following is 100% pure:

	100% pure?	
Freshly squeezed orange juice	Yes	No
Country air	Yes	No
Ocean water	Yes	No
Fresh drinking water	Yes	No
Skim milk	Yes	No
Stainless steel	Yes	No
A single water molecule	(Yes)	No

A glass of water contains on the order of a trillion trillion (1×10^{24}) molecules. If the water in this were 99.9999% pure, you could calculate the percent of impurities by subtracting from 100.0000%

$$100.0000\% \text{ water } + \text{ impurity molecules}$$
$$-\ 99.9999\% \text{ water molecules}$$
$$0.0001\% \text{ impurity molecules}$$

Pull out your calculator and calculate the number of impurity molecules in the glass of water. Do this by finding 0.0001% of 1×10^{24}, which is the same as multiplying 1×10^{24} by 0.000001.

$$(1 \times 10^{24})(0.000001) = \underline{1 \times 10^{18}}$$

How many impurity molecules are there in a glass of water that's 99.9999% pure?

a) 1,000 (one thousand: 10^3) b) 1,000,000 (one million: 10^6)

c) 1,000,000,000 (one billion: 10^9) (circled) 1,000,000,000,000,000,000 (one million trillion: 10^{18}).

How does your answer make you feel about drinking water that is 99.9999 percent free of some poison, such as a pesticide?
That there are a million trillion poison molecules in a glass of water might make one hesitate... but read on!

For every one impurity molecule, how many water molecules are there? (Divide the number of water molecules by the number of impurity molecules.)

$$10^{24}/10^{18} = 10^6 = 1,000,000 = \text{one million}$$

Would you describe these impurity molecules within water as "rare" or "common"? For every one impurity molecule there are one million water molecules. One in a million is rare!

A friend argues that he or she doesn't drink tap water because it contains thousands of molecules of some impurity in each glass. How would you respond in defense of the water's purity, if it indeed does contain thousands of molecules of some impurity per glass?
Only a 1,000 impurity molecules in this glass of water would make this water incredibly pure... about 99.9999999999999999 % pure!

CONCEPTUAL PHYSICAL SCIENCE EXPLORATIONS

Chapter 23 Chemical Bonding
Chemical Bonds

1. Based upon their positions in the periodic table, predict whether each pair of elements will form an ionic, covalent, or neither (atomic number in parenthesis)

a. Gold (79) and Platinum (78)   N
b. Rubidium (37) and Iodine (53)   I
c. Sulfur (16) and Chlorine (17)   C
d. Sulfur (16) and Magnesium (12)   I
e. Calcium (20) and Chlorine (17)   I
f. Germanium(32) and Arsenic (33)   C
g. Iron (26) and Chromium (24)   N
h. Chlorine (17) and Iodine (53)   C
i. Carbon (6) and Bromine (35)   C
j. Barium (56) and Astatine (85)   I

2. The most common ions of lithium, magnesium, aluminum, chlorine, oxygen, and nitrogen and their respective charges are as follows:

Positively Charged Ions
Lithium ion: Li^{1+}
Barium ion: Ba^{2+}
Aluminum ion: Al^{3+}

Negatively Charged Ions
Chloride ion: Cl^{1-}
Oxide ion: O^{2-}
Nitride ion: N^{3-}

Use this information to predict the chemical formulas for the following ionic compounds:

a. Lithium Chloride: $LiCl$
b. Barium Chloride: $BaCl_2$
c. Aluminum Chloride: $AlCl_3$
d. Lithium Oxide: Li_2O
e. Barium Oxide: BaO
f. Aluminum Oxide: Al_2O_3
g. Lithium Nitride: Li_3N
h. Barium Nitride: Ba_3N_2
i. Aluminum Nitride: AlN

j. How are elements that form positive ions grouped in the periodic table relative to elements that form negative ions? POSITIVE ION ELEMENTS TOWARD THE LEFT AND NEGATIVE IONS TOWARD THE RIGHT.

3. Predict whether the following chemical structures are polar or nonpolar:

POLAR POLAR

NONPOLAR NONPOLAR

NONPOLAR NONPOLAR

CONCEPTUAL PHYSICAL SCIENCE EXPLORATIONS

Chapter 23 Chemical Bonding
Shells and the Covalent Bond

When atoms bond covalently their atomic shells overlap so that shared electrons can occupy both shells at the same time.

Non-bonded hydrogen atoms

Hydrogen   Hydrogen

Covalently bonded hydrogen atoms

Molecular Hydrogen
Formula: H_2

Fill each shell model shown below with enough electrons to make each atom electrically neutral. Use arrows to represent electrons. Within the box draw a sketch showing how the two atoms bond covalently. Draw hydrogen shells more than once when necessary so that no electrons remain unpaired. Write the name and chemical formula for each compound.

A.

Hydrogen   Carbon

Name of Compound: **METHANE** Formula: CH_4

B.

Hydrogen   Nitrogen

Name of Compound: **AMMONIA** Formula: NH_3

C.

Hydrogen   Oxygen

Name of Compound: **WATER** Formula: H_2O

D.

Hydrogen   Fluorine

Name of Compound: **HYDROGEN FLUORIDE** Formula: **HF**

E.

Hydrogen   Neon

BONDING NOT POSSIBLE!

Name of Compound: _____ Formula: _____

1. Note the relative positions of carbon, nitrogen, oxygen, fluorine, and neon in the periodic table. How does this relate to the number of times each of these elements is able to bond with hydrogen?

IT'S IN A DESCENDING ORDER FROM LEFT TO RIGHT.

2. How many times is the element boron (atomic number 5) able to bond with hydrogen? Use the shell model to help you with your answer.

ONLY 3 VALENCE ELECTRONS, THEREFORE, ONLY 3 BONDS.

CONCEPTUAL PHYSICAL SCIENCE EXPLORATIONS

Chapter 23 Chemical Bonding
Bond Polarity and the Shell Model

"Pretend you are one of two electrons being shared by a hydrogen atom and a fluorine atom. Say, for the moment, you are centrally located between the two nuclei. You find that both nuclei are attracted to you, hence, because of your presence the two nuclei are held together."

You are here

H : F

1. Why are the nuclei of these atoms attracted to you? __BECAUSE OF YOUR NEGATIVE CHARGE__

2. What type of chemical bonding is this? __COVALENT__

You are held within hydrogen's 1st shell and at the same time within fluorine's 2nd shell. Draw a sketch using the shell models below to show how this is possible. Represent yourself and all other electrons usings arrows. Note your particular location.

Hydrogen   Fluorine

Your Sketch

3. You are pulled toward the hydrogen nucleus, which has a positive charge. How strong is this charge from your point of view—what is its *electronegativity?* $\sim +1$

4. You are also attracted to the fluorine nucleus. What is its *electronegativity?* $\sim +7$

"You are being shared by the hydrogen and fluorine nuclei. But as a moving electron you have some choice as to your location."

5. Consider the electronegativities you experience from both nuclei. Which nucleus would you tend to be closest to? __FLUORINE__

91

"Stop pretending you are an electron and observe the hydrogen-fluorine bond from outside the hydrogen fluoride molecule. Bonding electrons tend to congregate to one side because of the differences in effective nuclear charges. This makes one side slightly negative in character and the opposite side slightly positive. Indicate this on the following structure for hydrogen fluoride using the symbols δ^- and δ^+."

H : F

By convention, bonding electrons are not shown. Instead, a line is simply drawn connecting the two bonded atoms. Again indicate the slightly negative and positive ends.

H — F

6. Would you describe hydrogen fluoride as a polar or nonpolar molecule? __POLAR__

7. If two hydrogen fluoride molecules were thrown together would they stick or repel? (Hint: what happens when you throw two small magnets together?) __STICK__

8. Place bonds between the hydrogen and fluorine atoms to show many hydrogen fluoride molecules grouped together. Each element should be bonded only once. Circle each molecule and indicate the slightly negative and slightly positive ends.

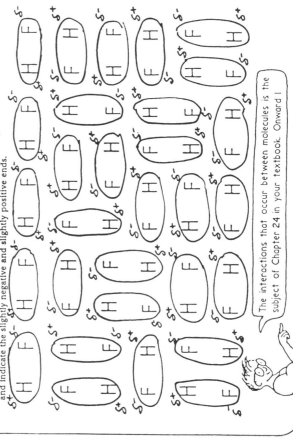

"The interactions that occur between molecules is the subject of Chapter 24 in your textbook. Onward !"

92

CONCEPTUAL PHYSICAL SCIENCE EXPLORATIONS

Chapter 24 Molecular Mixing
Solutions

1. Use these terms to complete the following sentences. Some terms may be used more than once.

solution	solvent	solute
dissolve	concentrated	dilute
saturated	concentration	mole
insoluble	solubility	soluble
	precipitate	supersaturated

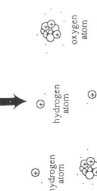

Sugar is __SOLUBLE__ in water for the two can be mixed homogeneously to form a __SOLUTION__.

The __SOLUBILITY__ of sugar in water is so great that __CONCENTRATED__ homogeneous mixtures are easily prepared. Sugar, however, is not infinitely __SOLUBLE__ in water for when too much of this __SOLUTE__ is added to water, which behaves as the __SOLVENT__, the solution becomes __SATURATED__. At this point any additional sugar is __INSOLUBLE__ for it will not __DISSOLVE__. If the temperature of a saturated sugar solution is lowered, the __SOLUBILITY__ of the sugar in water is also lowered. If some of the sugar comes out of solution, it is said to form a __PRECIPITATE__. If, however, the sugar remains in solution despite the decrease in solubility, then the solution is said to be __SUPER-SATURATED__. Adding only a small amount of sugar to water results in a __DILUTE__ solution. The __CONCENTRATION__ of this solution or any solution can be measure in terms of __MOLARITY__, which tells us the number of solute molecules per liter of solution. If there are 6.022×10^{23} molecules in 1 liter of solution, then the __CONCENTRATION__ of the solution is 1 __MOLE__ per liter.

2. Temperature has a variety of effects on the solubilites of various solutes. With some solutes, such as sugar, solubility increases with increasing temperature. With other solutes, such as sodium chloride (table salt), changing temperature has no significant effect. With some solutes, such as lithium sulfate, Li_2SO_4, the solubility actually decreases with increasing temperature.

a. Describe how you would prepare a supersaturated solution of lithium sulfate.

__FORM A SATURATED SOLUTION AND THEN SLOWLY RAISE THE TEMPERATURE.__

b. How might you cause a saturated solution of lithium sulfate to form a precipitate?

__INCREASE ITS TEMPERATURE__

CONCEPTUAL PHYSICAL SCIENCE EXPLORATIONS

Chapter 24 Molecular Mixing
Atoms to Molecules to Molecular Attractions

Subatomic particles are the fundamental building blocks of all __ATOMS__.

protons   neutrons   electrons

SUBATOMIC PARTICLES

An atom is a group of __SUBATOMIC PARTICLES__ held tightly together. An oxygen atom is a group of 8 __PROTONS__, 8 __NEUTRONS__, and 8 __ELECTRONS__. A hydrogen atom is a group of only 1 __PROTON__ and 1 __ELECTRON__.

hydrogen atom

hydrogen atom   oxygen atom

oxygen atom   hydrogen atom

ATOMS

A __MOLECULE__ is a group of atoms held tightly together. A water __MOLECULE__ consists of 2 __OXYGEN__ atoms and 1 __HYDROGEN__ atom.

water molecule   water molecule

MOLECULES

Water is a material made up of billions upon billions of water __MOLECULES__. The physical properties of water are based upon how these water __MOLECULES__ interact with one another. The electronic attractions between __MOLECULES__ is the main topic of Chapter 24.

WATER

CONCEPTUAL PHYSICAL SCIENCE EXPLORATIONS

Chapter 26 Oxidation and Reduction

Loss and Gain of Electrons

A chemical reaction that involves the transfer of an electron is classified as an oxidation-reduction reaction. Oxidation is the process of losing an electrons, while reduction is the process of gaining them. Any chemical that causes another chemical to lose electrons (become oxidized) is called an oxidizing agent. Conversely, any chemical that causes another chemical to gain electrons is called a reducing agent.

1. What is the relationship between an atom's ability to behave as an oxidizing agent and its electron affinity?

THE GREATER THE ELECTRON AFFINITY, THE GREATER ITS ABILITY TO BEHAVE AS AN OXIDIZING AGENT.

2. Relative to the periodic table, which elements tend to behave as strong oxidizing agents?

THOSE TO THE UPPER LEFT WITH THE EXCEPTION OF THE NOBLE GASES.

3. Why don't the noble gases behave as oxidizing agents?

THEY HAVE NO SPACE IN THEIR SHELLS TO ACCOMODATE ADDITIONAL ELECTRONS.

4. How is it that an oxidizing agent is itself reduced?

REDUCTION IS THE GAINING OF ELECTRONS. IN PULLING AN ELECTRON AWAY FROM ANOTHER ATOM, AN OXIDIZING AGENT NECESSARILY GAINS AN ELECTRON.

5. Is it possible to have an endothermic oxidation-reduction reaction? If so, cite examples.

YES ... THE ELECTROLYSIS OF WATER IS AN ENDOTHERMIC OXIDATION-REDUCTION REACTION.

6. Specify whether each reactant is about to be oxidized or reduced.

$$2\ K\ \ +\ \ H_2O\ \longrightarrow\ 2\ K^+\ +\ {}^-OH$$
OX   RED

$$2\ Mg\ +\ O_2\ \longrightarrow\ 2\ Mg^{2+}\ O^{2-}$$
OX   RED

$$2\ Na\ +\ Cl_2\ \longrightarrow\ 2\ Na^+\ Cl^-$$
OX   RED

$$CH_4\ +\ 2\ O_2\ \longrightarrow\ O=C=O\ +\ H-O-H$$
OX   RED

7. Which oxygen atom enjoys a greater negative charge?

$O=O$ — this one   -or- $H-O-H$ (that one)   (circle one)

8. Relate your answer to question 7 to how it is that O_2 is reduced upon reacting with CH_4 to form carbon dioxide and water. IN TRANSFORMING FROM O_2 TO H_2O, AN OXYGEN ATOM IS GAINING ELECTRONS IS BEST AS IT CAN. WITH ITS GREATER NEGATIVE CHARGE IT CAN BE THOUGHT OF "AS REDUCED".

Donating and Accepting Hydrogen Ions continued:

We see from the previous reaction that because the ammonium ion donated a hydrogen ion, it behaved as an acid. Conversely, the hydroxide ion by accepting a hydrogen ion behaved as a base. How do the ammonia and water molecules behave during the reverse process?

$$[H_3N-H]^+\ +\ {}^-O-H\ \longrightarrow\ H_3N\ +\ H-O-H$$
acid   base   BASE   ACID
                  ammonia   water

Identify the following molecules as behaving as an acid or a base:

$$HO-P(=O)(OH)-O^-\ +\ {}^-O-H\ \rightleftharpoons\ HO-P(=O)(O^-)-O^-\ +\ H_2O^+-H$$
ACID   BASE   BASE   ACID

$$H_3N-H\ +\ NaH\ \rightleftharpoons\ [H_3N-H]^+\ Na^+\ +\ H-H$$
ACID   BASE   BASE   ACID

$$H-H\ +\ {}^-H\ \rightleftharpoons\ H^-\ +\ H-H$$
ACID   BASE   BASE   ACID

$$HNO_3\ +\ NH_3\ \rightleftharpoons\ {}^-NO_3\ +\ {}^+NH_4$$
ACID   BASE   BASE   ACID

CONCEPTUAL PHYSICAL SCIENCE EXPLORATIONS

Chapter 28 Chemistry of Drugs
Neurotransmitters

Many drugs work by mimicking neurotransmitters. Amphetamine, for example, mimicks the stimulatory neurotransmitter norepinephrine. Much of this occurs in the synaptic cleft, which is a narrow gap between neurons.

Stick Structures:

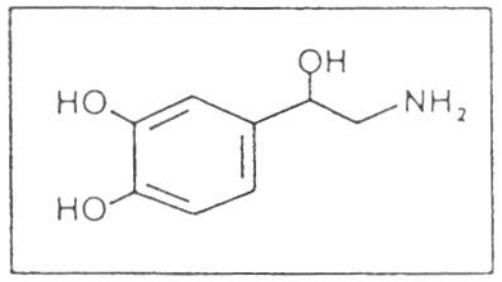

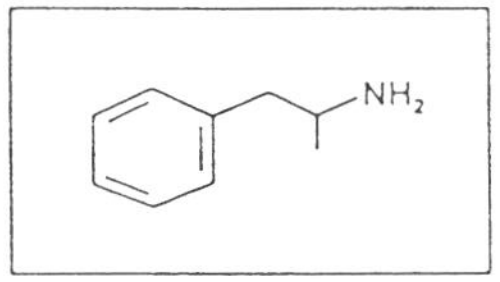

Norepinephrine

Amphetamine

Schematic Representation:

1. Under normal conditions, the pre-synaptic neuron releases norepinephrine, which migrates across the synaptic cleft to a receptor site on the surface of the post-synaptic neuron.

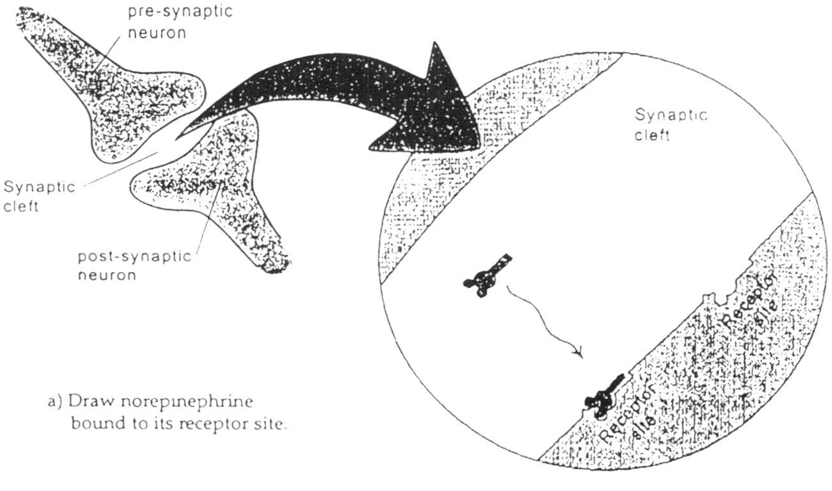

pre-synaptic neuron

Synaptic cleft

post-synaptic neuron

Synaptic cleft

a) Draw norepinephrine bound to its receptor site.

CONCEPTUAL PHYSICAL SCIENCE EXPLORATIONS

Chapter 27 Organic Compounds
Structures of Organic Compounds

1. What are the chemical formulas for the following structures.

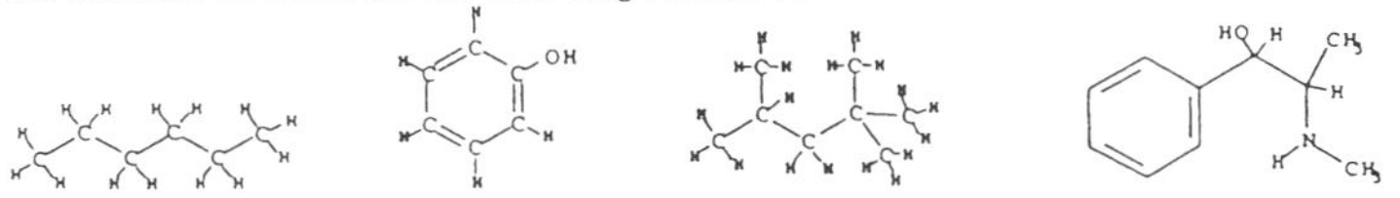

Formula: C_6H_{14} CH_6O C_8H_{18}

2. How many covalent bonds is carbon able to form? __4__

3. What is wrong with the structure shown in the box:

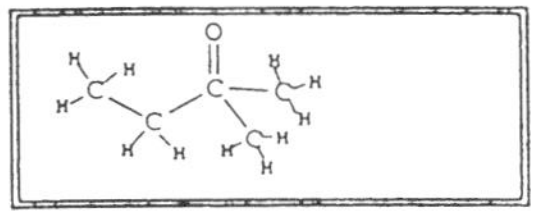

THE CARBON OF THE CARBONYL IS BONDED 5 TIMES

$C_{10}H_{15}NO$

4a. Draw a hydrocarbon that contains 4 carbon atoms.

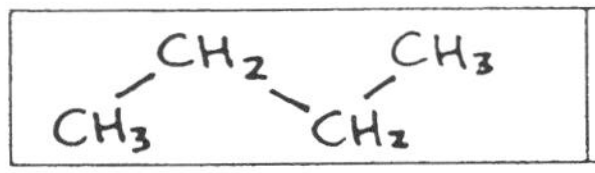

4b. Redraw your structure and transform it into an amine.

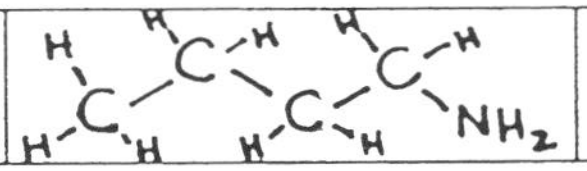

4c. Transform your amine into an amide. You may need to relocate the nitrogen.

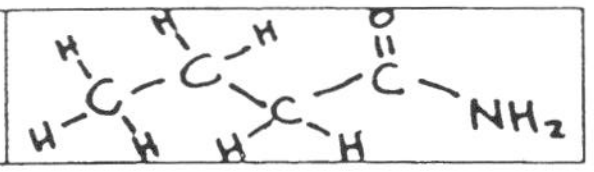

4d. Redraw your amide transforming it into a carboxylic acid.

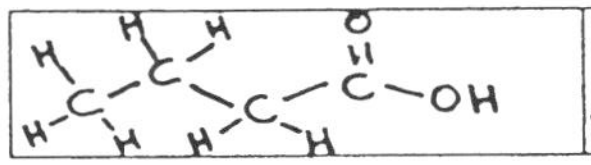

4e. Redraw your carboxylic acid transforming it into an alcohol.

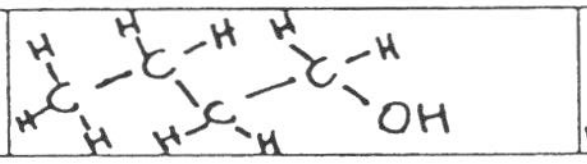

4f. Rearrange the carbons of your alcohol to make an ether.

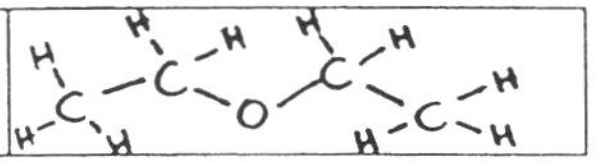

5. Circle the following alkaloids that are in their free-base form?

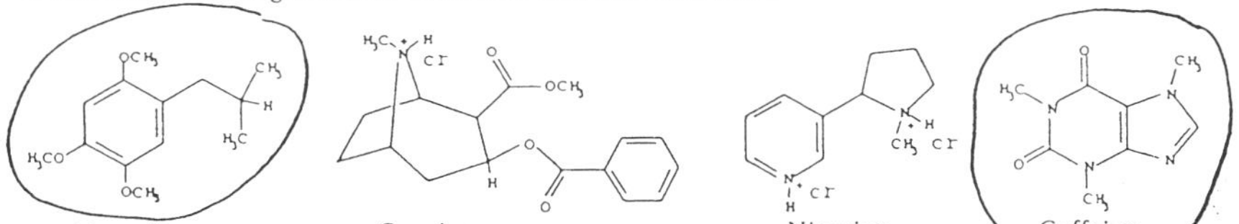

Mescaline

Cocaine

Caffeine

Nicotine

6. How might you convert a free-base alkaloid into a salt?
REACT IT WITH HYDROCHLORIC ACID

7. Why are alkaloids less water soluble in their free-base form?
THEY ARE FAR LESS POLAR

8. Which should have a greater tendency to vaporize upon heating: an alkaloid in its free-base form or one in its salt form? Why? FREE BASE. ELECTRICAL INTERACTIONS BETWEEN
THE IONS OF THE SALT ARE MUCH GREATER

CONCEPTUAL PHYSICAL SCIENCE EXPLORATIONS

Chapter 29 Plastics
Polymers

1. Circle the monomers that may be useful for forming an addition polymer and draw a box around the ones that may be useful for forming a condensation polymer.

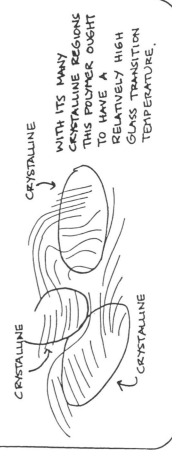

2. Which type of polymer always weighs less than the sum of its parts? Why?

THE CONDENSATION POLYMERS LOSE SMALL MOLECULES SUCH AS WATER WHEN THEY FORM AND THUS THE POLYMER THAT FORMS WEIGHS LESS THAN THE SUM OF ITS MONOMERS.

3. Would a material with the following arrangement of polymer molecules have a relatively high or low melting point? Why?

WITH ITS MANY CRYSTALLINE REGIONS THIS POLYMER OUGHT TO HAVE A RELATIVELY HIGH GLASS TRANSITION TEMPERATURE.

CRYSTALLINE

CRYSTALLINE

CRYSTALLINE

105

---

2. When amphetamine molecules get within the cleft, there is too much stimulation of the post-synaptic neuron.

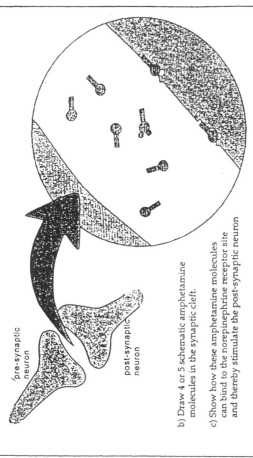

pre-synaptic neuron

post-synaptic neuron

b) Draw 4 or 5 schematic amphetamine molecules in the synaptic cleft.

c) Show how these amphetamine molecules can bind to the norepinephrine receptor site and thereby stimulate the post-synaptic neuron

3. In response to the overstimulation, the body stops producing so much norepinephrine. Also, the amphetamine molecules are eventually broken down and the result is shown below:

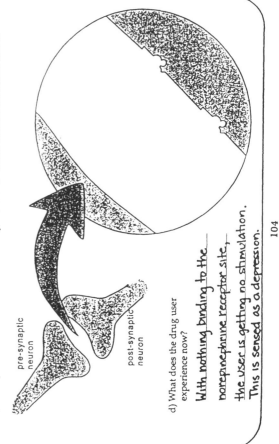

pre-synaptic neuron

post-synaptic neuron

d) What does the drug user experience now?

With nothing binding to the norepinephrine receptor site, the user is getting no stimulation. This is sensed as a depression.

104

CONCEPTUAL PHYSICAL SCIENCE **EXPLORATIONS**

Chapter 30 Minerals and Their Formation
Chemical Structure and Formulas of Minerals

Out of the more than 3400 minerals, only about two dozen are abundant. Minerals are classified by their chemical composition and internal atomic structure and are divided into groups. For this exercise we explore the following mineral groups: *carbonates, sulfides, sulfates,* and *halides.*

For each mineral structure diagrammed below, look for a pattern in the structure, count the number of atoms (ions) in each, and fill in the blanks.

> The schematic diagrams are simple representations of small mineral structures. Actual mineral structures extend farther and comprise more atoms.

1. Circle pairs of Na and Cl ions in the structure and add any ion(s) needed to complete pairing. This mineral structure contains __14__ Na ions and __14__ Cl ions. The mineral's formula is __NaCl__. This mineral belongs to the __Halide__ group.

⊘ Cl   ⊙ Na

2. This mineral structure contains __8__ Ca atoms, __8__ C atoms, and __24__ O atoms. The mineral's formula is __CaCO3__. This mineral belongs to the __CARBONATE__ group.

◯ Ca   • C   ○ O

Chapter 29 Plastics
History of Plastics

Turn the page sideways. Read the descriptions of plastics and using Sections 29.3 – 29.6 of your textbook, identify the name of each plastic for that time. Write its name in the blank that shows the correct time of its discovery.

> For example, write rubber vulcanization in the blank for 1839.

1839 Discovered by the addition of sulfur to a natural product.

1845 Discovered by the wiping of a spill of nitric and sulfuric acids.

1855 Formed when nitrocellulose was treated with solvents, such as alcohol.

1870 Formed when nitrocellulose was treated with camphor.

1892 Made by treating cellulose with harsh chemicals.

1907 The first hardset plastic widely used a molding medium.

1913 Viscose manufactured into a thin, transparent sheet.

1926 Thin, transparent sheet of viscose with small amounts of nitrocellulose.

1929 First polymer that was made workable by the addition of a plasticizer.

1930 Allowed the United States to make tires after entering World War II.

1935 Developed in Britain and used to coat wiring for the first RADAR devices.

1937 First used for ladies hosiery prior to being used for making military parachutes.

1938 A moldable plastic glass useful for manufacturer of gunner's ports in airplanes.

1938 A polymer used in the development of the atomic bomb.

1940 Originally designed to protect theater seats from chewing gum.

1950 Introduced as the first synthetic substitute for wool.

Timeline (1830 – 1950):

- Vulcanized rubber — 1839
- Nitrocellulose/gun cotton
- Collodion — 1855
- Celluloid — 1870
- Viscose/Rayon — 1892
- Cellophane — 1913
- Vaporproof cellophane — 1907
- Bakelite — 1907
- Synthetic Rubber — 1930
- Polyethylene — 1933
- Nylon — 1937
- Saran — 1940
- Teflon — 1938
- PVC — 1929
- Plexiglas — 1938
- Dacron polyester — 1950

Many specialized polymers have been produced since the 1950s.

Timeline years: 1830 1840 1850 1860 1870 1880 1890 1900 1910 1920 1930 1940 1950

CONCEPTUAL PHYSICAL SCIENCE **EXPLORATIONS**

Chapter 31 Rocks
The Rock Cycle

Complete the illustration, which depicts the different paths in the rock cycle. Insert arrows to show direction of pathways.

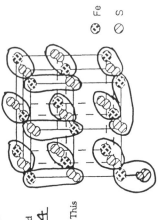

IGNEOUS ROCK

Magma

Metamorphic Rock

Sediment

SEDIMENTARY ROCK

Weathering, erosion, transportation, deposition

Lithification

1. Can a rock that has undergone metamorphism turn into sedimentary rock? If so, how? If not, how come?

Yes, metamorphic rock subjected to weathering breaks down into sediment. As the weathered material undergoes lithification it becomes sedimentary rock.

2. By what process does hot molten magma become rock?

Hot molten magma becomes igneous rock after it has cooled and solidified (the process of crystallization).

3. List three rocks generated from the different types of magma.

Basalt, Andesite, and Granite (others too!)

4. Carbonate rocks found in the Colorado Rocky Mountains imply what type of deposition environment?

A shallow sea environment, the remains of ancient sea floors.

5. In what section(s) of the rock cycle are gemstones formed?

All sections. Gem forms from the slow cooling of magma-peridot and topaz form in igneous environments. Pressure and heat result in the recrystallization of minerals-metamorphism. Garnet is a metamorphic gem. Only a few gems, such as turquoise and opal, are actually formed in a sedimentary environment.

6. The Big Island of Hawaii is almost 1 million years old. Yet hikers there almost never step on rock that is more than 1 thousand years old. Explain.

The island of Hawaii is still in the process of formation. The island is a large shield volcano formed from the accumulation of successive lava flows. With each new flow, rocks from previous flows are covered. So the rock we walk on is never more than a thousand years old.

---

3. This mineral structure contains ___6___ Ca atoms, ___6___ S atoms, and ___24___ O ions. The mineral's formula is __CaSO4__ This mineral belongs to the __SULFATE__ group.

○ Ca
⊘ S
○ O

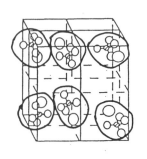

⊙ Fe
⊘ S

4. Complete the structure by adding the needed atom(s). This mineral structure contains ___14___ Fe atoms and ___26___ S atoms. The mineral's formula is __FeS2__. This mineral belongs to the __SULFITE__ group.

5. Complete the mineral structure so that each Ca atom is linked to two F atoms. Now the mineral structure contains ___14___ F atoms. The Ca atoms and ___28___ F atoms. The mineral's formula is __CaF2__. This mineral belongs to the __HALIDE__ group.

○ Ca
⊙ F

y
Name _____ Period _____ Date _____

CONCEPTUAL PHYSICAL SCIENCE EXPLORATIONS

Chapter 32 The Architecture of the Earth
Faults

Three block diagrams are illustrated below. Draw arrows on each diagram to show the direction of movement. Answer the questions next to each diagram.

A.

What type of force produced Fault A?

__TENSIONAL__

Name the fault __NORMAL FAULT__

Where would you expect to find this type of fault?

__NEVADA, PARTS OF CA, OR, IDAHO__
__AND UTAH__

B.

What type of force produced Fault B?

__COMPRESSIONAL__

Name the fault __REVERSE OR THRUST FAULT__

Where would you expect to find this type of fault?

__ROCKY MTN FORELAND,__
__APPALACHIAN MTNS__

C.

What type of force produced Fault C?

__SHEAR__

Name the fault __HORIZONTAL MOVEMENT FAULT__

Where would you expect to find this type of fault?

__CALIFORNIA — THE SAN ANDREAS__
__FAULT__

111

Chapter 31 Rocks
Igneous Rock Differentiation: How to Make Granite

A mineral is called a high temperature mineral if its melting/freezing temperature is relatively high. A mineral is called a low temperature mineral if its melting/freezing temperature is relatively low.

Suppose we start with solid, basaltic rock. If it is heated, it will partially melt.

1. Is the type of mineral left behind (that doesn't liquefy) a high temperature or low temperature mineral?

__High temperature mineral__   (Do you think granite could form in this manner?)

2. Which type will melt to form a liquid?

__Low temperature__

3. Will the resulting liquid be higher or lower in silicon content than the original rock? Why?
__Higher. Low temperature minerals have a higher silica content__
__than high temperature minerals.__

4. If this liquid is separated from the original rock and then cooled relatively quickly, what is the name of the rock that will most likely form?

__Andesite__

5. Repeat steps 1 through 4 for the rock formed in question 4. What is the name of the resulting rock if the liquid is allowed to cool very slowly?

__Granite__

Now consider a magma chamber that contains completely molten basaltic magma. Let's allow this magma to cool very slowly.

6. Which type of minerals will be the first to form, low temperature or high temperature minerals?

__High temperature minerals__

7. Will the remaining liquid be higher or lower in silicon content than the original liquid? Why?
__Higher — high temperature minerals are lower in silica content__
__so the liquid becomes enriched in silica.__

Assume that the newly formed crystals settle to the bottom of the magma chamber so that there is no chemical interaction between the newly formed crystals and the remaining liquid.

8. If this process continues, will the low temperature minerals eventually crystallize?

__Yes__

9. If so, would a rock formed from these minerals be higher or lower in silicon content than a basalt?

__It will be higher.__

110

197

CONCEPTUAL PHYSICAL SCIENCE EXPLORATIONS

Chapter 32 The Architecture of the Earth
Structural Geology

Much subsurface information is learned by oil companies when wells are drilled. Some of this information leads to the discovery of oil, and some reveals subsurface structures such as folds and/or faults in the Earth's crust.

Four oil wells that have been drilled to the same depth are shown on the cross section below. Each well encounters contacts between different rock formations at the depths shown in the table below. Rock formations are labeled A — F, with A as the youngest rock formation and F as the oldest rock formation.

Contact	Depth to Contact (in meters)			
	Oil well #1	Oil well #2	Oil well #3	Oil well #4
A-B	200	not encountered	200	not encountered
B-C	400	100	400	100
C-D	600	300	600	300
D-E	800	500	800	500
E-F	1000	700	1000	700

1. In the cross section below, Contacts D - E and E - F are plotted for Oil Wells 1 and 2. Plot the remainder of the data for all four wells, labeling each point you plot.

2. Draw lines to connect the contacts between the rock formations (as is done for Contacts D - E and E - F for Oil Wells 1 and 2).

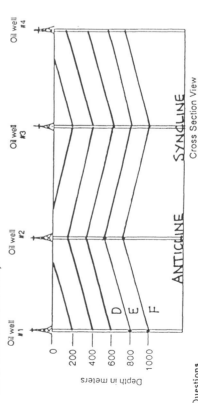

ANTICLINE   SYNCLINE

Cross Section View

Questions

1. What explanation can you offer for no sign of formation A in Wells 2 and 4?

FORMATION A HAS BEEN ERODED AWAY.

2. What geological structures are revealed? Label them on the cross section.
THE STRUCTURES ARE FOLDED. AN ANTICLINE IS EVIDENT AT WELL #2, A SYNCLINE AT WELL #3.

113

CONCEPTUAL PHYSICAL SCIENCE EXPLORATIONS

Chapter 33 Our Restless Planet
Plate Boundaries

Draw arrows on the plate boundaries A, B, and C, to show the relative direction of movement.

Type of boundary for A? **CONVERGENT**
What type of force generates this type of boundary?

COMPRESSIONAL

Is this a site of lithospheric formation, destruction, or lithospheric transport?

DESTRUCTION

Type of boundary for B? **TRANSFORM FAULT**
What type of force generates this type of boundary?

SHEAR

Is this a site of lithospheric formation, destruction, or lithospheric transport?

TRANSPORT

Type of boundary for C? **DIVERGENT**
What type of force generates this type of boundary?

TENSIONAL

Is this a site of lithospheric formation, destruction, or lithospheric transport?

FORMATION

Draw arrows on the transform faults to indicate relative motion.

— Transform fault

|||| Mid-ocean spreading ridge

115

Chapter 33 Our Restless Planet
Sea-Floor Spreading

The rate of sea floor spreading, some 1 to 10 centimeters per year, is found by knowing the distance and age between two points on the ocean floor. Diagrams A, B, and C show stages of sea-floor spreading. Spreading begins at A, continues to B where rocks at locations P begin to spread to the farther-apart positions we see in C. At C newer rock at the ocean crest S is dated at 10 million years.

Using the scale: 1mm = 50km, use a ruler on C to find:

1. The separation rate of the two continental landmasses in the past 10 million years, in cm/yr $\underline{5}$

2. The age of the sea floor at P in diagram C (1cm/yr = 10km/million years) $\underline{95\ MILLION\ YEARS}$

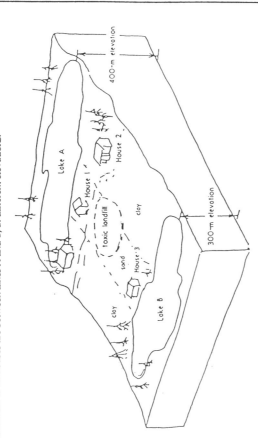

Sea level

A

B

C

CALCULATIONS FOR SEA FLOOR SPREADING

1) S to ridge crest = 5mm = 250 km

$$\frac{250\ km}{10\ m.y.} = 25\ km/m.y.$$

$$\left(\frac{25\ km}{m.y.}\right)\left(\frac{10^5 cm}{1\ km}\right)\left(\frac{1\ m.y.}{10^5\ y}\right) = 2.5\ cm/yr$$

2) P to ridge crest = 47.5mm = 2375 km

$$\frac{2375\ km}{25\ km/m.y.} = 95\ m.y.$$

$$\frac{2375\ km}{95\ m.y.} = 25\ km/m.y.$$

Continental crust · Old oceanic crust · New oceanic crust · Lithosphere · Asthenosphere · Sediments

---

CONCEPTUAL PHYSICAL SCIENCE **EXPLORATIONS**
Chapter 34 Water on our World
Groundwater Flow and Contaminant Transport

The occupants of Houses 1, 2, and 3 wish to drill wells for domestic water supply. Note that the locations of all houses are between lakes A and B, at different elevations.

400-m elevation

Lake A

House 1

House 2

toxic landfill

clay

sand

House 3

clay

Lake B

clay

300-m elevation

1. Show by sketching dashed lines on the drawing, the likely direction of groundwater flow beneath all the houses.
   See drawing

2. Which of the wells drilled beside Houses 1, 2, and 3 are likely to yield an abundant water supply?
   Wells at houses 1 and 3 should yield a sufficient amount of water because sand is quite permeable. The well at house 2 is in clay which has low permeability and so will not yield a sufficient water supply.

3. Do any of the three need to worry about the toxic landfill contaminating their water supply? Explain.
   House 2 doesn't have a decent water supply—no contamination worries, House 1 is upgradient from the landfill so will have no contamination worries unless pumping rate is high enough to draw water against the gradient. House 3 is in big trouble.

4. Why don't the homeowners simply take water directly from the lakes?
   Sand acts as a good filter for bacteria and viruses. Also, the additional residence time in groundwater allows chemical reactions to remove many contaminants.

5. Suggest a potentially better location for the landfill. Defend your choice.
   A better location would be in clay. Clay's low permeability would hinder leaching of contaminants to the groundwater.

CONCEPTUAL PHYSICAL SCIENCE EXPLORATIONS

Chapter 35 Our Natural Landscape
Stream Flow

The diagram below illustrates a stream. Deposition of sediment occurs in one area, and erosion of sediment occurs in another area. On the diagram below mark the areas of deposition and the areas of erosion.

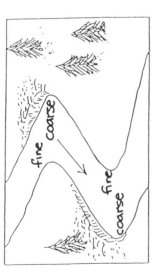

Farther downstream, the shape of the stream channel changes. Draw a likely new shape in the box below.

(ANY STREAM THAT INCREASES IN SINUOSITY)

Place these words where they belong in the blanks below:
(coarse-grained increase decreases fine-grained)

As the stream continues to meander, it widens the stream's valley into a broad, low-lying area called a *floodplain*. Floodplains are so named because they are the sites of periodic flooding. In a flood, as discharge and flow speed **INCREASE**, so does the stream's ability to carry sediment. So, when a stream overflows its banks, sediment-rich water spills out onto the floodplain. Because the speed of the water quickly **DECREASES** as it spreads out over the large, flat, floodplain, a sequence of coarse to fine particles is deposited. Along the edges of the channel we can expect to find **COARSE-GRAINED** sediment. Farther away from the stream channel, on the floodplain, we can expect to find **FINE-GRAINED** sediments.

119

Chapter 35 Our Natural Landscape
Stream Velocity

Let's explore how the average velocity of streams and rivers can change. Recall in Chapter 35 that the volume of water that flows past a given location over any given length of time depends both on the stream velocity and the cross-sectional area of the stream. We say

$$Q = A \times V$$

where Q is the volumetric flow rate, A is the cross sectional area of the stream, and V its average velocity.

Consider the stream shown below, with rectangular cross sectional areas

$A = $ width $\times$ depth

> V is the average velocity. For rivers, flow is usually a bit faster just below the surface and a bit slower along the bottom. So, strictly speaking, velocity varies with depth.

(diagram showing Location 1 with A_1, and Location 2 with A_2, each labeled depth and width)

1. The two locations shown have no inlets or outlets between them, so Q remains constant. Suppose the cross-sectional areas are also constant ($A_1 = A_2$), with Location 2 deeper but narrower than Location 1. What change, if any, occurs for the stream velocity?

There is no change in average velocity

2. If Q remains constant, what happens to stream velocity at Location 2 if A_2 is less than A_1?

Average velocity increases at location 2.

3. If Q remains constant, what happens to stream velocity at Location 2 if A_2 is greater than A_1?

Average velocity decreases at Location 2.

4. What happens to stream velocity at Location 2 if area A_2 remains the same, but Q increases (perhaps by an inlet along the way?)

Average velocity increases at Location 2.

5. What happens to stream velocity at Location 2 if both A_2 and Q increase?

It depends. If Q increases more than A increases, average velocity increases. If A increases more than Q increases average velocity decreases. If they both increase at the same proportion, there is no change in average velocity.

120

CONCEPTUAL PHYSICAL SCIENCE EXPLORATIONS

Chapter 36 A Brief History of the Earth
Relative Time — What Came First?

The cross section below depicts many geologic events. In the space to the right, list the different events starting with the oldest to the youngest event. Where appropriate, include tectonic events (such as folding, deposition of beds, subsidence, uplift, erosion, intrusion).

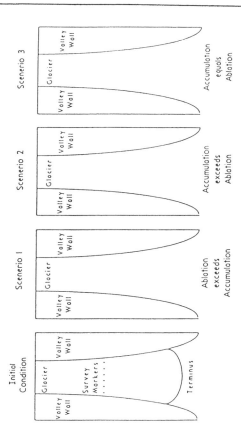

Youngest

P deposition
   erosion
H faulting
G
E subsidence
   deposition
E uplift
Q erosion
N
M subsidence
   deposition
L uplift
   erosion
   faulting
I intrusion
   folding
K
J
D
C
   subsidence
B deposition
Oldest A nonconformity

Examine the rings in the cross section of a tree and you do more than determine the age of the tree. Relative thicknesses of the rings tells a lot about the climate conditions throughout the tree's history. A geologist similarly learns much about the Earth's history by examination of rock layers in cross sections of the Earth's crust.

CONCEPTUAL PHYSICAL SCIENCE EXPLORATIONS

Chapter 35 Our Natural Landscape
Glacial Movement

From season to season the mass of a glacier changes. With each change in mass, the glacier moves. Glacier movement is measured by placing a line of markers across the ice and recording their changes in position over a period of time.

In the example below, we show the initial position of a line of markers and the glaciers terminus (the end of the glacier).

Draw the markers and glacier terminus at a later time for each of the following scenarios.

Initial Condition

Valley Wall   Glacier   Valley Wall
Survey Markers . . . .
Valley Wall   Terminus

Scenerio 1

Valley Wall   Glacier   Valley Wall

Ablation exceeds Accumulation

Scenerio 2

Valley Wall   Glacier   Valley Wall

Accumulation exceeds Ablation

Scenerio 3

Valley Wall   Glacier   Valley Wall

Accumulation equals Ablation

1. Define accumulation.   The annual amount of snow added to a glacier.

2. Define ablation.   The annual amount of ice lost by a glacier.

3. Which scenario shows the greatest glacial mass?   Scenerio 2

4. In which part of the glacier does ice move the fastest?   In the center

5. In which part of the glacier does ice move the slowest? Why?   Along the edges—the sides and bottom of a glacier are impacted by frictional drag.

6. For each scenario, what, if anything, is the same?   In all three scenarios the glacier is moving downslope.

CONCEPTUAL PHYSICAL SCIENCE EXPLORATIONS

Chapter 36 A Brief History of the Earth
Unconformities and Age Relationships

The wavy lines in the 4 regions below represent unconformities. Investigate the regions and answer corresponding questions about the 4 regions below.

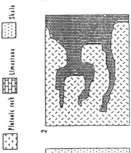
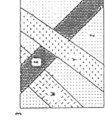
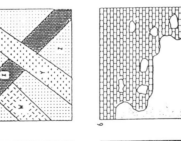
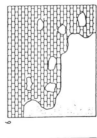

1. Did the faulting and dike occur before or after the unconformity? __before__
   What kind of unconformity is it? __nonconformity__

2. Did the faulting occur before or after the unconformity? __after__

3. Did the folding occur before or after the unconformity? __before__
   What kind of unconformity is it? __angular unconformity__

4. What kind of unconformity is represented? __angular unconformity__

5. Interestingly, the age of the Earth is some 4.5 billion years old—yet the oldest rocks found are some 3.8 billion years old. Why do we find no 4.5-billion year old rocks?
   The Earth's earliest crustal surface has been reworked. Rocks that formed 4.5 b.y.a. have been remelted into new rock.

6. What is the age of the innermost ring in a living redwood tree that is 2000 years old? What is the age of the outermost ring? How does this example relate to the previous question?
   The innermost ring is 2000 years old; the outermost ring is 1 year old. New layers cover older layers.

7. What is the approximate age of the atoms that make up a 3.8-billion year old rock?
   They are older than 3.7 billion years.

125

Chapter 36 A Brief History of the Earth
Age Relationships

From your investigation of the 6 geologic regions shown, answer the questions below.
The number of each question refers to the same-numbered region.

Granite | Basalt | Metallic rock | Limestone | Shale | Sandstone

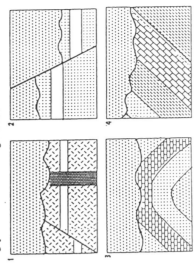

1. The shale has been cut by a dike. The radiometric age of the dike is estimated at 40 million years. Is the shale younger or older? __older__

2. Which is older, the granite or the basalt? __the granite__

3. The sandstone bed, Z, has been intruded by dikes. What is the age succession of dikes, going from oldest to youngest? __W, X, Y__

4. Which is older, the shale or the basalt? __the shale__

5. Which is older, the sandstone or the basalt? __the basalt__

6. Which is older, the sandstone or the limestone? __the sandstone__

124

CONCEPTUAL PHYSICAL SCIENCE EXPLORATIONS

Chapter 36 A Brief History of the Earth
Radiometric Dating

Isotopes Commonly Used for Radiometric Dating

Radioactive Parent	Stable Daughter Product	Half-life Value
Uranium-238	lead-206	4.5 billion years
Uranium-235	lead-207	704 million years
Potassium-40	argon-40	1.3 billion years
Carbon-14	nitrogen-14	5730 years

1. Consider a radiometric lab experiment wherein 99.98791 % of a certain radioactive sample of material remains after one year. What is the decay rate of the sample?

$$1.000... - 0.9998791 = 0.0001209 \text{ /yr}$$

2. What is the rate constant?
(Assume that the decay rate is constant for the one year period.)

$$K = \frac{\text{decay rate}}{\text{starting amount}} = \frac{0.0001209/\text{yr}}{1.000...} = 0.0001209/\text{yr}$$

3. What is the half-life?

$$T = \frac{0.693}{0.0001209/\text{yr}} = 5732 \text{ yr}$$

4. Identify the isotope.

Carbon 14

5. In a sample collected in the field, this isotope was found to be 1/16 of its original amount. What is the age of the sample?

Note from graph that it is 4 half-lives $4 \times 5732 = 22928$ yr

You need to know:
• Decay rate = (amount decayed) / time
• Rate constant K = (decay rate/ starting amount)
  in units 1/year
• Half-life T = 0.693/K (units in years)

[graph: Proportion of atoms remaining vs Time, in half-lives]

Chapter 36 A Brief History of the Earth
Our Earth's Hot Interior

A major puzzle faced scientist in the 19th century. Volcanoes showed that the Earth's interior is semi-molten. Penetration into the crust by bore-holes and mines showed that the Earth's temperature increases with depth. Scientists knew that heat flows from the interior to the surface. They assumed that the source of the Earth's internal heat was primordial, the afterglow of its fiery birth. Measurements of the Earth's rate of cooling indicated a relatively young Earth—some 25 to 30 million years in age. But geological evidence indicated an older Earth. This puzzle wasn't solved until the discovery of radioactivity. Then it was learned that the interior was kept hot by the energy of radioactive decay. We now know the age of the Earth is some 4.5 billion years—a much older Earth.

All rock contains trace amounts of radioactive minerals. Radioactive minerals in common granite release energy at the rate 0.03 J/kg·yr. Granite at the Earth's surface transfers this energy to the surroundings practically as fast as it is generated, so we don't find granite any warmer than other parts of our environment. But what if a sample of granite were thermally insulated? That is, suppose all the increase of thermal energy due to radioactive decay were contained. Then it would get hotter. How much hotter? Let's figure it out, using 790 J/kg·C° as the specific heat of granite.

Let's see now... back in Chapter 9 we learned that the relationship between quantity of heat, mass, specific heat and temperature difference is
$$Q = cm\Delta T$$

Calculations to make:

1. How many joules are required to increase the temperature of 1 kg of granite by 500 C°?

$$Q = cm\Delta T$$
$$= (1\text{kg})(790\text{ J/kg·C°}) 500\text{ C°} = 395,000 \text{ J}$$

2. How many years would it take radioactivity in a kilogram of granite to produce this many joules?

$$\frac{395,000 \text{ J}}{0.03 \text{ J/kg·yr}} \times 1\text{ kg} \approx 13 \text{ million years}$$

Questions to answer:

1. How many years would it take a thermally insulated 1-kilogram chunk off granite to undergo a 500 C° increase in temperature?

Same   13 million years

2. How many years would it take a thermally insulated one-million-kilogram chunk off granite to undergo a 500 C° increase in temperature?

Same (correspondingly more radiation!)

3. Why does the Earth's interior remain molten hot?

Because of radioactivity

4. Rock has a higher melting temperature deep in the interior. Why?

Greater pressure (like water in a pressure cooker)

An electric toaster stays hot while electric energy is supplied, and doesn't cool until switched off. Similarly, do you think the energy source now keeping the Earth hot will one day suddenly switch off like a disconnected toaster—gradually decrease over a long time?

5. Why doesn't the Earth just keep getting hotter until it all melts?

Interior is not perfectly insulated - heat migrates to surface

CONCEPTUAL PHYSICAL SCIENCE **EXPLORATIONS**

Chapter 37 The Atmosphere, The Oceans, and Their Interactions
The Earth's Seasons

1. The warmth of equatorial regions and coldness of polar regions on the Earth can be understood by considering light from a flashlight striking a surface. If it strikes perpendicularly, light energy is more concentrated as it covers a smaller area; if it strikes at an angle, the energy spreads over a larger area. So the energy per unit area is less.

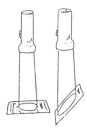

The arrows represent rays of light from the distant sun incident upon the Earth. Two areas of equal size are shown, Area A near the north pole and Area B near the equator. Count the rays that each reach each area, and explain why region B is warmer than region A.

3 RAYS INCIDENT ON A; 6 ON B. SO REGION B GETS TWICE AS MUCH SOLAR ENERGY AND IS WARMER.

2. The Earth's seasons result from the 23.5-degree tilt of the Earth's daily spin axis as it orbits the sun. When the Earth is at the position shown on the right in the sketch below (not shown to scale), the Northern Hemisphere tilts toward the sun, and sunlight striking it is strong (more rays per area). Sunlight striking the Southern Hemisphere is weak (fewer rays per area). Days in the north are warmer, and daylight lasts longer. You can see this by imagining the Earth making its complete daily 24-hour spin.

Do two things on the sketch: (1) Shade the Earth in nighttime darkness for all positions, as is already done in the left position. (2) Label each position with the proper month — March, June, September, or December.

(DEC) (SEPT) (MAR) (JUNE)

Be sure to do the shading before you answer the questions on the backside of this sheet!

a. When the Earth is in any of the four positions shown, during one 24-hour spin a location at the equator receives sunlight half the time and is in darkness the other half of the time. This means that regions at the equator always get about __12__ hours of sunlight and __12__ hours of darkness.

b. Can you see that in the June position regions farther north have longer days and shorter nights? Locations north of the Arctic Circle (dotted in the Northern Hemisphere) are always illuminated by the sun as the Earth spins, so they get daylight __24__ hours a day.

c. How many hours of light and darkness are there in June at regions south of the Antarctic Circle (dotted line in Southern Hemisphere)?
ZERO HOURS OF LIGHT, OR 24 HOURS OF DARKNESS PER DAY

d. Six months later, when the Earth is at the December position, is the situation in the Antarctic the same or is it the reverse?
REVERSE; MORE SUNLIGHT PER AREA IN DEC IN SOUTHERN HEMISPHERE.

e. Why do South America and Australia enjoy warm weather in December instead of in June?
IN DEC THE SOUTHERN HEMISPHERE TILTS TOWARD THE SUN AND GETS MORE SUNLIGHT PER AREA THAN IN JUNE.

3. The Earth spins about its polar axis once each 24 hours, which gives us day and night. If the Earth's spin was instead only one rotation per year, what difference would there be with day and night as we enjoy them now?
ONE FACE OF THE EARTH WOULD ALWAYS BE IN SUNLIGHT, AND THE OPPOSITE SIDE WOULD ALWAYS BE IN DARKNESS.

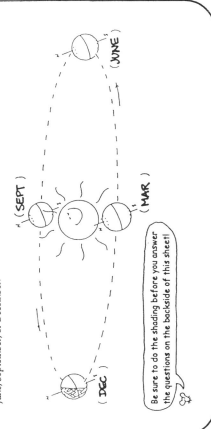

If the spin of the Earth was the same as its revolution rate around the sun, would we be like the moon — one side always facing the body it orbits?

In Section 39.4 read ahead about gravity lock and why the moon shows only one face to Earth.

CONCEPTUAL PHYSICAL SCIENCE **EXPLORATIONS**

Chapter 37 The Atmosphere, The Oceans, and Their Interactions
Short and Long Wavelengths

The sine curve is a pictorial representation of a wave—the high points being crests, and the low points troughs. The height of the wave is its *amplitude*. The wavelength is the distance between successive identical parts of the wave (like between crest to crest, or trough to trough). Wavelengths of water waves at the beach are measured in meters, wavelengths of ripples in a pond are measured in centimeters, and the wavelengths of light in billionths of a meter (nanometers).

Crest ——— wavelength ——— crest

amplitude

wavelength

trough

trough

In the boxes below sketch three waves of the same amplitude—Wave A with half the wavelength of Wave B, and Wave C with wavelength twice as long as Wave B.

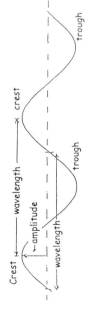

Wave A

Wave B

Wave C

1. If all three waves have the same speed, which has the highest frequency? __Wave A__

2. Compared with solar radiation, terrestrial radiation has a __long__ wavelength.

3. In a florist's greenhouse, __short__ waves are able to penetrate the greenhouse glass, but __long__ waves cannot.

4. The Earth's atmosphere is similar to the glass in a greenhouse. If the atmosphere were to contain excess amounts of water vapor and carbon dioxide, the air would be opaque to __long__ waves.

Chapter 37 The Atmosphere, The Oceans, and Their Interactions
Driving Forces of Air Motion

The primary driving force of the Earth's weather is __sunlight__. The unequal distribution of solar radiation on the Earth's surface creates temperature differences which in turn result in pressure differences in the atmosphere. These pressure differences generate horizontal winds as air moves from __high__ pressure to __low__ pressure. The weather patterns are not strictly horizontal though, there are other forces affecting the movement of air. Recall from Newton's second law that an object moves in the direction of the *net* force acting on it. The forces acting on the movement of air include:

1) pressure gradient force 2) Coriolis force 3) centripetal force, and 4) friction

The greater the pressure difference the greater the force, and the greater the wind. The "push" caused by the horizontal differences in pressure across a surface is called the *pressure gradient force*. This force is represented by isobars on a weather map. Isobars connect locations on a map that have equal atmospheric pressure. The pressure gradient force is perpendicular to the isobars and strongest where the isobars are closely spaced. So,

the steeper the pressure gradient, the __stronger__ the wind.

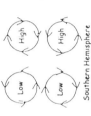

Low gradient, weak winds

1020
1030
1040
1050
H

Steep gradient, strong winds

The *Coriolis force* is a result of the Earth's rotation. The Coriolis force is the deflection of the wind from a __straight__ path to a __curved__ path. The Coriolis force causes the wind to veer to the right of its path in the Northern Hemisphere and to the left of its path in the Southern Hemisphere.

N

Equator

S

As the wind blows around a low or high pressure center it constantly changes its direction. A change in speed or direction is acceleration. In order to keep the wind moving in a circular path the net force must be directed inward. This __inward__ force is called *centripetal force*.

The forces described above greatly influence the flow of upper winds (winds not influenced by surface frictional forces). The interaction of these forces cause the winds in the Northern Hemisphere to rotate __clockwise__ around regions of high pressure and __counter-clockwise__ around regions of low pressure. In the Southern Hemisphere the situation is reversed — winds rotate __clockwise__ around a high and __clockwise__ around a low.

Northern Hemisphere

High   High   Low   Low

Southern Hemisphere

Winds blowing near the Earth's surface are slowed by *frictional forces*. In the Northern Hemisphere surface winds blow in a direction __counter-clockwise__ into the centers of a low pressure area and __clockwise__ out of the centers of a high pressure area. The spiral direction is reversed in the Southern Hemisphere. Draw arrows to show the direction of the pressure gradient force.

High

Northern Hemisphere

Low

High

Southern Hemisphere

205

CONCEPTUAL PHYSICAL SCIENCE **EXPLORATIONS**

Chapter 38 Weather
Air Temperature and Pressure Patterns

Temperature patterns on weather maps are depicted by Isotherms—lines that connect all points having the same temperature. Each isotherm separates temperatures having higher values from temperatures having lower values.

The following weather map to the right shows temperatures in degrees Fahrenheit for various locations. Using 10 degree intervals, connect same value numbers to construct isotherms. Label the temperature value at each end of the isotherm. One isotherm has been completed as an example.

Tips for drawing Isotherms
- Isotherms can never be open ended.
- Isotherms are "closed" if they reach the boundary of plotted data, or make a loop.
- Isotherms can never touch, cross, or fork.
- Isotherms must always appear in sequence; for example, there must be a 60° isotherm between a 50- and 70-degree isotherm.
- Isotherms should be labeled with their values.

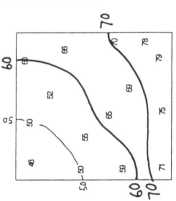

Pressure patterns on weather maps are depicted by isobars—lines which connect all points having equal pressure. Each isobar separates stations of higher pressure from stations of lower pressure.

The weather map below shows air pressure in millibar (mb) units at various locations. Using an interval of **4** (for example, 1008, 1012, 1016 etc.), connect equal pressure values to construct isobars. Label the pressure value at each end of the isobar. One isobar has been completed as an example.

- Tips for drawing isobars are similar to those for drawing isotherms.

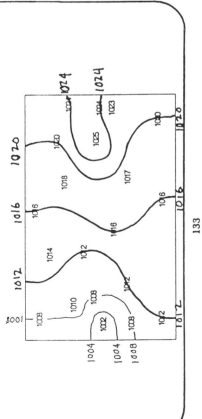

Air Temperature and Pressure Patterns continued:

On the map above, use an interval of 4 to draw lines of equal pressure (isobars) to show the pattern of air pressure. Locate and mark regions of high pressure with an "H" and regions of low pressure with an "L". Use the map to answer the questions below.

1. On the map above, areas of high pressure are depicted by the ___1024___ isobar.

2. On the map above, areas of low pressure are depicted by the ___1008___ isobar.

Circle the correct answer

3. Highs are usually accompanied by (stormy weather) (fair weather).

4. In the Northern Hemisphere, surface winds surrounding a high pressure system blow in a (clockwise direction) (counterclockwise direction).

5. In the Northern Hemisphere, surface winds spiral inward into a (region of low pressure) (region of high pressure).

CONCEPTUAL PHYSICAL SCIENCE EXPLORATIONS
Chapter 38: Weather
Surface Weather Maps

Station models are used on weather maps to depict weather conditions for individual localities. Weather codes are plotted in, on, and around a central circle that describes the overall appearance of the sky. Jutting from the circle is a wind arrow, its tail in the direction from which the wind comes and its feathers indicating the wind speed. Other weather codes are in standard position around the circle.

Use the simplified station model and weather symbols to complete the statements below.

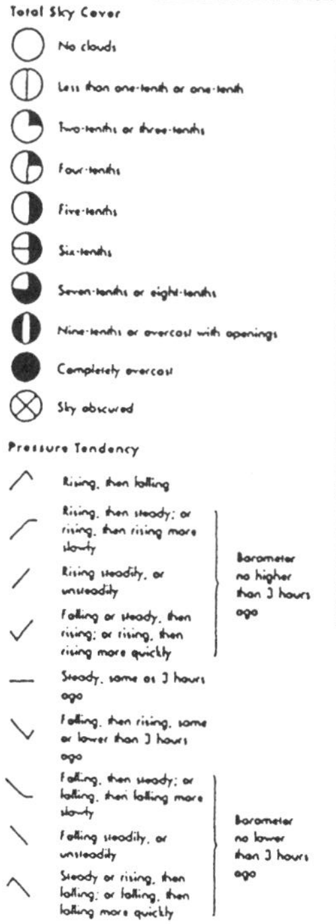

Wind Entries

	Miles (Statute) Per Hour	Knots	Kilometers Per Hour
Calm	Calm	Calm	Calm
	1-2	1-2	1-3
	3-8	3-7	4-13
	9-14	8-12	14-19
	15-20	13-17	20-32
	21-25	18-22	33-40
	26-31	23-27	41-50
	32-37	28-32	51-60
	38-43	33-37	61-69
	44-49	38-42	70-79
	50-54	43-47	80-87
	55-60	48-52	88-96
	61-66	53-57	97-106
	67-71	58-62	107-114
	72-77	63-67	115-124
	78-83	68-72	125-143

Total Sky Cover
No clouds
Less than one-tenth or one-tenth
Two-tenths or three-tenths
Four-tenths
Five-tenths
Six-tenths
Seven-tenths or eight-tenths
Nine-tenths or overcast with openings
Completely overcast
Sky obscured

Pressure Tendency
Rising, then falling
Rising, then steady; or rising, then rising more slowly — Barometer no higher than 3 hours ago
Rising steadily, or unsteadily
Falling or steady, then rising; or rising, then rising more quickly
Steady, same as 3 hours ago
Falling, then rising, same or lower than 3 hours ago — Barometer no lower than 3 hours ago
Falling, then steady; or falling, then falling more slowly
Falling steadily, or unsteadily
Steady or rising, then falling; or falling, then falling more quickly

Common Weather Symbols
Light rain • ᐧ Rain shower
Moderate rain ᐧᐧ Snow shower
Heavy rain Showers of hail
Light snow Drizzling or blowing snow
Moderate snow Dust storm
Heavy snow = Fog
Light drizzle ∞ Haze
Ice pellets (sleet) Smoke
Freezing rain Thunderstorm
Freezing drizzle Hurricane

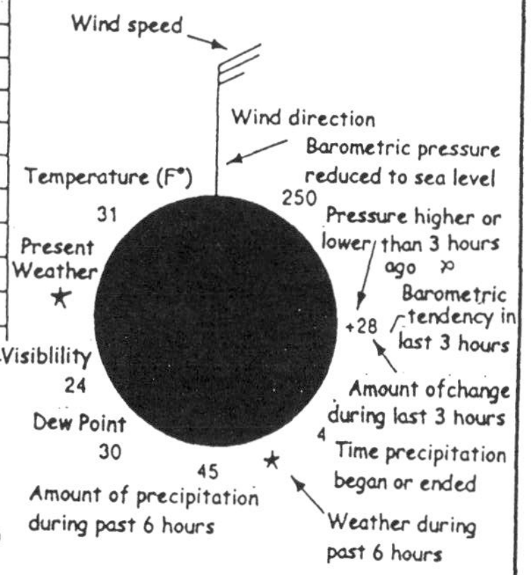

Station Model labels:
Wind direction
Wind speed
Temperature (F°) 31
Present Weather
Visibility 24
Dew Point 30
Amount of precipitation during past 6 hours 45
Barometric pressure reduced to sea level 250
Pressure higher or lower than 3 hours ago +28
Barometric tendency in last 3 hours
Amount of change during last 3 hours
Time precipitation began or ended
Weather during past 6 hours

1. The overall appearance of the sky is __completely overcast__
2. The wind speed is __41 – 50__ kilometers per hour.
3. The wind direction is coming from the __East.__
4. The present weather conditions call for __snow__
5. The barometric tendency is __rising, then steady__
6. For the past 6 hours the weather conditions have been __snowy.__

Surface Weather Maps continued:

Use the unlabeled station model to answer the questions below.

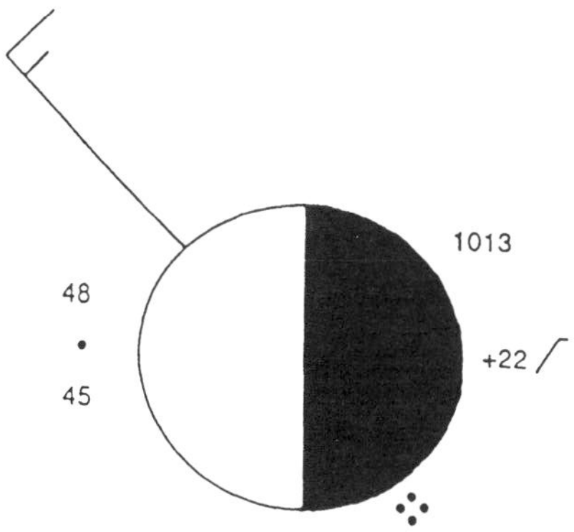

(Station model: 1013, +22, 48, 45)

1. The overall appearance of the sky is __five-tenths overcast__
2. The wind speed is __20 – 32__ kilometers per hour.
3. The wind direction is coming from the __East.__
4. The present weather conditions call for __light rain.__
5. The barometric tendency is __rising, then steady.__
6. For the past 6 hours the weather conditions have been __heavy rain.__
7. The barometric pressure is __1013 mb__
8. The dew point is __45__
9. The current temperature is __48°F__
10. Compared to the past few hours, barometric pressure is __increasing__

CONCEPTUAL PHYSICAL SCIENCE **EXPLORATIONS**

Chapter 39 The Solar System
Earth-Moon-Sun Alignments

You cast a shadow whenever you stand in the sunlight. Everything does, including planetary bodies. To better understand this, consider the sketch of the shadow cast by the apple. Note how the rays define the the darkest part of the shadow, the *umbra*, and the lighter part, the *penumbra*. During a solar eclipse, the moon similarly casts a shadow on the Earth. The region of "totality" is the umbra, and regions getting a partial eclipse are in the penumbra.

Below is a diagram of the sun, Earth, and the orbital path of the moon (dashed circle). One position of the moon is shown. Draw the moon in the appropriate positions on the dashed circle to represent (a) a quarter moon; (b) a half moon; (c) a solar eclipse; (d) a lunar eclipse. Label your positions. For c and d, extend rays from the top and bottom of the sun to show umbra and penumbra regions on the Earth.

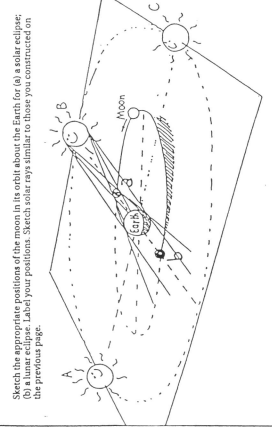

Eclipses are relatively rare because the orbital planes of the Earth about the sun and the moon about the Earth are slightly tipped to each other — so the shadows usually "miss." If the planes weren't tipped, eclipses would occur monthly. Can you see that eclipses only occur when the three bodies align along the intersection of these planes (points A and B)?

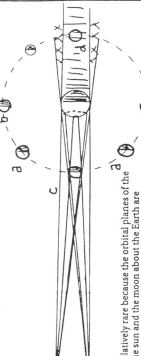

Because of the large size of the sun, its rays taper as shown. Note that the shadow of the new moon misses the Earth, and the shadow of the Earth misses the full moon. Would these shadows miss when alignment of the three bodies is along points A and B? __NO__

137

Earth-Moon-Sun Alignment continued:

Sketch the appropriate positions of the moon in its orbit about the Earth for (a) a solar eclipse; (b) a lunar eclipse. Label your positions. Sketch solar rays similar to those you constructed on the previous page.

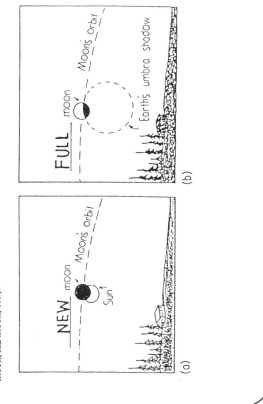

Shown below are (a) a partial solar eclipse in progress, and (b) a partial lunar eclipse in progress. Fill in the blanks and label the correct phases of the moon at these times (New moon, quarter moon, full moon, etc.).

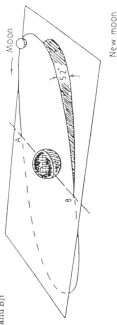

(a) NEW moon — Moon's orbit — Sun

(b) FULL moon — Moon's orbit — Earth's umbra shadow

138

CONCEPTUAL PHYSICAL SCIENCE EXPLORATIONS

Chapter 40 The Stars
Stellar Parallax

Finding distances to objects beyond the solar system is based on the simple phenomenon of parallax. Hold a pencil at arm's length and view it against a distant background — each eye sees a different view (try it and see). The displaced view indicates distance. Likewise, when the Earth travels around the sun each year, the position of relatively nearby stars shifts slightly relative to the background stars. By carefully measuring this shift, astronomer types can determine the distance to nearby stars.

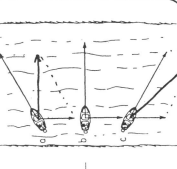

Background stars

Close star

Angle of parallax shift

Earth

Can you see why the close star appears to shift positions relative to the background stars? And how maximum shift appears in observations 6 months apart?

The photographs below show the same section of the evening sky taken at a 6-month interval. Investigate the photos carefully and determine which star appears in a different position in observations 6 months apart. Circle the star that shows a parallax shift.

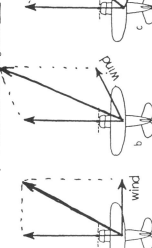

A

B

Below are three sets of photographs, all taken at 6-month intervals. Circle the stars that show a parallax shift in each of the photos.

Set A

Set B

Set C

Use a fine ruler and measure the distance of shift in millimeters and place the values below:

Set A **21** mm   Set B **7**   Set C **14** mm

Which set of photos indicate the closest star? The most distant "parallaxed" star?

A: GREATER PARALLAX.   B: MOST DISTANT

139

CONCEPTUAL PHYSICAL SCIENCE EXPLORATIONS

Appendix C Vectors
Vectors and the Parallelogram Rule

1. When vectors A and B are at an angle to each other, they add to produce the resultant C by the *parallelogram rule*. Note that C is the diagonal of a parallelogram where A and B are adjacent sides. Resultant C is shown in the first two diagrams, *a* and *b*. Construct the resultant C in diagrams *c* and *d*. Note that in diagram *d* you form a rectangle (a special case of a parallelogram).

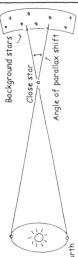

a

b

c

d

2. Below we see a top view of an airplane being blown offcourse by wind in various directions. Use the parallelogram rule to show the resulting speed and direction of travel for each case. In which case does the airplane travel fastest across the ground? **d** Slowest? **c**

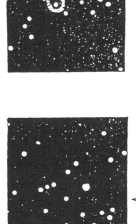

a

b

c

d

wind

wind

wind

wind

3. To the right we see top views of 3 motorboats crossing a river. All have the same speed relative to the water, and all experience the same water flow.

Construct resultant vectors showing the speed and direction of the boats.

a. Which boat takes the shortest path to the opposite shore? **a**

b. Which boat reaches the opposite shore first? **b**

c. Which boat provides the fastest ride? **c**

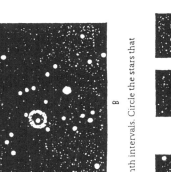

a

b

c

141

Appendix C Vectors
Velocity Vectors and Components

1. Draw the resultants of the four sets of vectors below.

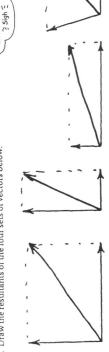

I was only a scalar until you came along and gave me direction! ? Sigh ?

2. Draw the horizontal and vertical components of the four vectors below.

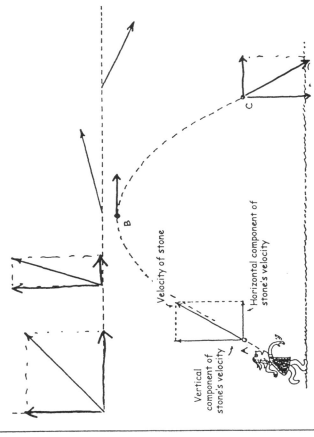

Velocity of stone

Vertical component of stone's velocity

Horizontal component of stone's velocity

3. She tosses the ball along the dashed path. The velocity vector, complete with its horizontal and vertical components, is shown at position A. Carefully sketch the appropriate velocity vectors with appropriate components for positions B and C.

a. Since there is no acceleration in the horizontal direction, how does the horizontal component of velocity compare for positions A, B, and C? __SAME__

b. What is the value of the vertical component of velocity at position B? __0 m/s__

c. How does the vertical component of velocity at position C compare with that of position A? __EQUAL AND OPPOSITE__

CONCEPTUAL PHYSICAL SCIENCE EXPLORATIONS

Appendix C Vectors
Vectors and Sailboats

(Do not attempt this until you have studied Appendix C!)

1. The sketch shows a top view of a small railroad car pulled by a rope. The force F that the rope exerts on the car has one component along the track, and another component perpendicular to the track.

a. Draw these components on the sketch. Which component is larger? __PERPENDICULAR COMPONENT__

b. Which component produces acceleration? __COMP PARALLEL TO TRACK__

c. What would be the effect of pulling on the rope if it were perpendicular to the track? __NO ACCELERATION__

2. The sketches below represent simplified top views of sailboats in a cross-wind direction. The impact of the wind produces a FORCE vector on each as shown. (We do NOT consider *velocity* vectors here!)

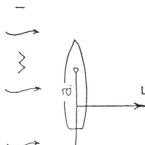

W I N D

a. Why is the position of the sail above useless for propelling the boat along its forward direction? (Relate this to Question 1c. above. Where the train is constrained by tracks to move in one direction, the boat is similarly constrained to move along one direction by its deep vertical fin — the *keel*.) __AS IN 1.C. ABOVE, THERE'S NO COMP PARALLEL TO DIRECTION OF MOTION.__

b. Sketch the component of force parallel to the direction of the boat's motion (along its keel), and the component perpendicular to its motion. Will the boat move in a forward direction? (Relate this to Question 1b. above.) __YES, AS IN 1.b. ABOVE, THERE IS A COMP PARALLEL TO DIRECTION OF MOTION.__

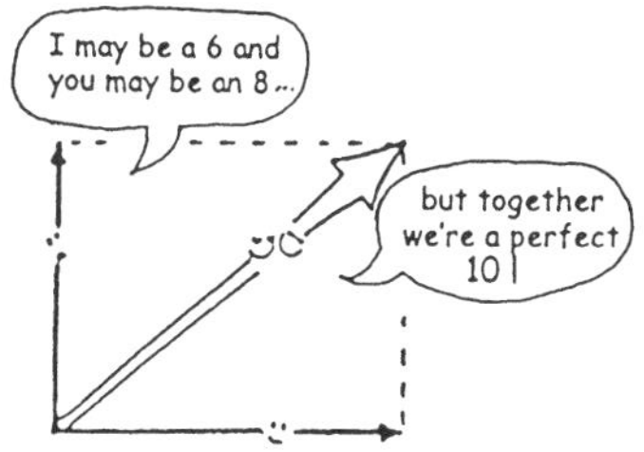

CONCEPTUAL PHYSICAL SCIENCE **EXPLORATIONS**
Appendix C Vectors
Force-Vector Diagrams

Being able to identify forces that act on a body is enormously important to any further study of physics. For example, in sketch 1 below we see two forces acting on the suspended rock — the force of gravity downward (W) and tension of the string (T) upward. These are the only forces that act on the rock. We see the rock in sketch 2 being pulled by two strings, so there are three forces, A, B, and C. The size of vectors A and B, relative to W, can be determined by the parallelogram rule. See if you can apply it to sketches 3 and 4. Continue identifying and sketching proper force vectors for the others. Good Energy!

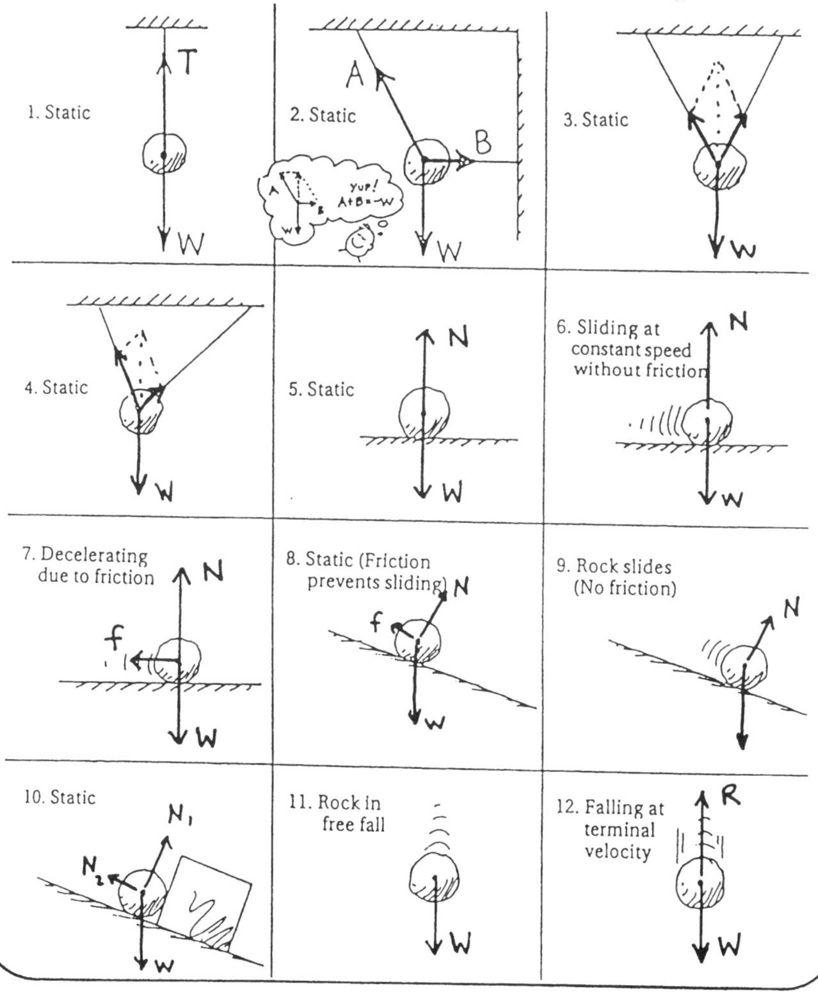

1. Static	2. Static	3. Static
4. Static	5. Static	6. Sliding at constant speed without friction
7. Decelerating due to friction	8. Static (Friction prevents sliding)	9. Rock slides (No friction)
10. Static	11. Rock in free fall	12. Falling at terminal velocity

145

---

3. The boat to the right is oriented at an angle into the wind. Draw the force vector and its forward and perpendicular components.

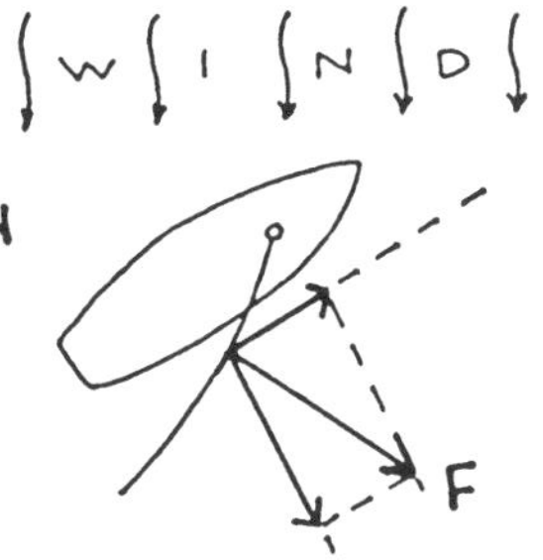

   a. Will the boat move in a forward direction and tack into the wind? Why or why not?

   YES, BECAUSE THERE IS A COMPONENT OF FORCE PARALLEL TO THE DIRECTION OF MOTION.

4. The sketch below is a top view of five identical sailboats. Where they exist, draw force vectors to represent wind impact on the sails. Then draw components parallel and perpendicular to the keels of each boat.

   a. Which boat will sail the fastest in a forward direction?

   BOAT 4 (WILL USUALLY EXCEED BOAT 1)

   b. Which will respond least to the wind?

   BOAT 2 (OR BOAT 3) *

   c. Which will move in a backward direction?

   BOAT 5

   d. Which will experience less and less wind impact with increasing speed?

   BOAT 1 (NO IMPACT AT WIND SPEED)

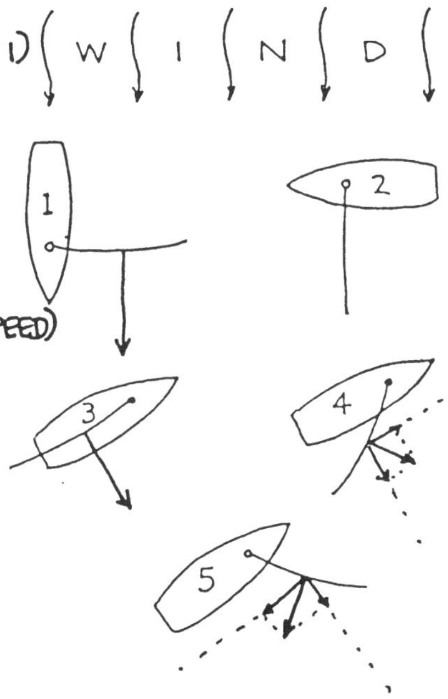

* THE WIND MISSES THE SAIL OF BOAT 2, AND THERE'S NO COMPONENT PARALLEL TO THE KEEL FOR BOAT 3.

I may be a 6 and you may be an 8...

but together we're a perfect 10!

CONCEPTUAL PHYSICAL SCIENCE EXPLORATIONS

Appendix D Fluid Physics
Archimedes' Principle I

1. Consider a balloon filled with 1 liter of water (1000 cm³) in equilibrium in a container of water, as shown in Figure 1.

WATER DOES NOT SINK IN WATER!

Figure 1

a. What is the mass of the 1 liter of water?

1 kg

b. What is the weight of the 1 liter of water?

9.8 N (or 10 N)

c. What is the weight of water displaced by the balloon?

9.8 N (or 10 N)

d. What is the buoyant force on the balloon?

9.8 N (or 10 N)

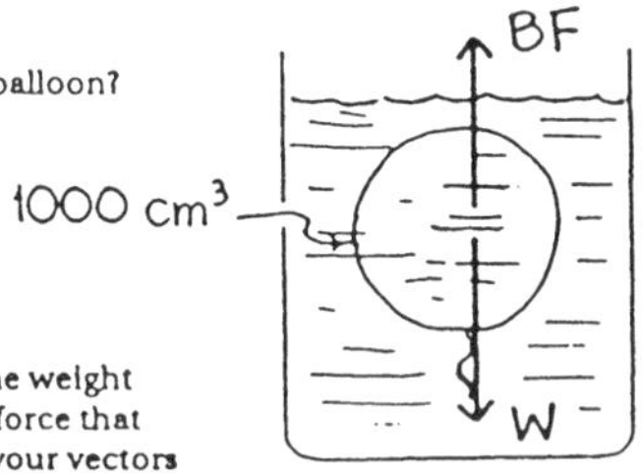

e. Sketch a pair of vectors in Figure 1: one for the weight of the balloon and the other for the buoyant force that acts on it. How do the size and directions of your vectors compare?

SAME SIZE, OPPOSITE DIRECTIONS

2. As a thought experiment, pretend we could remove the water from the balloon but still remain the same size of 1 liter. Then inside the balloon is a vacuum.

ANYTHING THAT DISPLACES 9.8 N OF WATER EXPERIENCES 9.8 N OF BUOYANT FORCE.

a. What is the mass of the liter of nothing?

0 kg

b. What is the weight of the liter of nothing?

0 N

c. What is the weight of water displaced by the massless balloon?

9.8 N (or 10 N)

d. What is the buoyant force on the massless balloon?

9.8 N (or 10 N)

CUZ IF YOU PUSH 9.8 N OF WATER ASIDE THE WATER PUSHES BACK ON YOU WITH 9.8 N !

e. In which direction would the massless balloon be accelerated?

UPWARD

147

Archimedes' Principle I continued:

3. Assume the balloon is replaced by a 0.5-kilogram piece of wood that has exactly the same volume (1000 cm³), as shown in Figure 2. The wood is held in the same submerged position beneath the surface of the water.

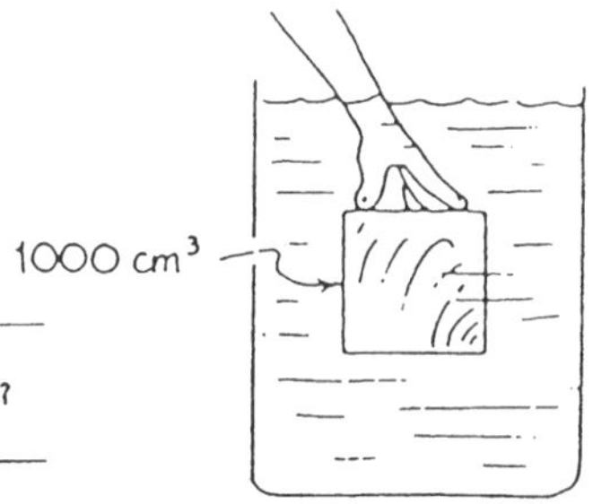

Figure 2

a. What volume of water is displaced by the wood?

1000 cm³ = 1 L

b. What is the mass of the water displaced by the wood?

1 kg

c. What is the weight of the water displaced by the wood?

9.8 N

d. How much buoyant force does the surrounding water exert on the wood?

9.8 N

e. When the hand is removed, what is the net force on the wood?

NET FORCE = BUOYANT FORCE – WEIGHT OF WOOD = 9.8 N – 4.9 N = 4.9 N (UPWARD)

THE BUOYANT FORCE ON A SUBMERGED OBJECT EQUALS THE WEIGHT OF WATER DISPLACED

... NOT THE WEIGHT OF THE OBJECT ITSELF!

... UNLESS IT IS FLOATING !

f. In which direction does the wood accelerate when released? **UPWARD**

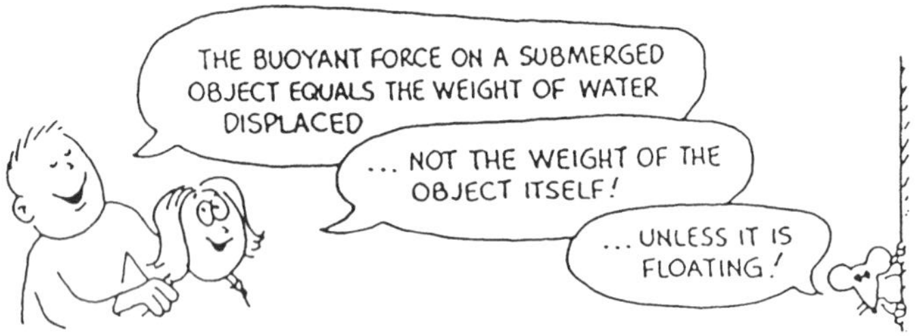

4. Repeat parts *a* through *f* in the previous question for a 5-kg rock that has the same volume (1000 cm³), as shown in Figure 3. Assume the rock is suspended by a string in the container of water.

Figure 3

a. **1000 cm³ (SAME)**

b. **1 kg (SAME)**

c. **9.8 N (SAME)**

d. **9.8 N (SAME)**

e. **39 N DOWNWARD***

f. **DOWNWARD**

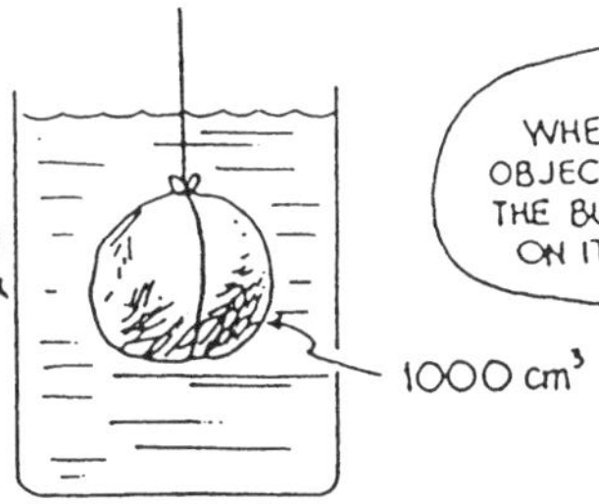

WHEN THE WEIGHT OF AN OBJECT IS GREATER THAN THE BUOYANT FORCE EXERTED ON IT, IT SINKS!

(DOWNWARD)

* NET FORCE = BUOYANT FORCE – WT ROCK = 9.8 N – 49 N = –39 N

148

CONCEPTUAL PHYSICAL SCIENCE EXPLORATIONS

Appendix D Fluid Physics
Archimedes' Principle II

1. The water lines for the first three cases are shown. Sketch in the appropriate water lines for cases *d* and *e*, and make up your own for case *f*.

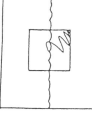

a. DENSER THAN WATER

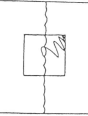

b. SAME DENSITY AS WATER

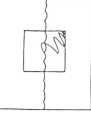

c. 1/2 AS DENSE AS WATER

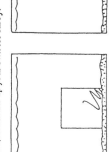

d. 1/4 AS DENSE AS WATER

e. 3/4 AS DENSE AS WATER

f. _____ AS DENSE AS WATER

(OPEN)

2. If the weight of a ship is 100 million N, then the water it displaces weighs **100 MILLION N.**
If cargo weighing 1000 N is put on board then the ship will sink down until an extra **1000 N** of water is displaced.

3. The first two sketches below show the water line for an empty and a loaded ship. Draw in the appropriate water line for the third sketch.

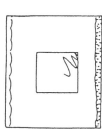

a. SHIP EMPTY

b. SHIP LOADED WITH 50 TONS OF IRON

c. SHIP LOADED WITH 50 TONS OF STYROFOAM

SAME!

Archimedes' Principle II continued:

4. Here is a glass of ice water with an ice cube floating in it. Draw the water line after the ice cube melts. (Will the water line rise, fall, or remain the same?)

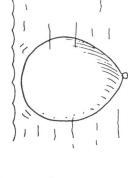

REMAINS SAME. VOL OF WATER WITH SAME WT OF ICE CUBE EQUALS VOL OF SUBMERGED PORTION OF ICE CUBE. THIS IS ALSO THE VOL OF WATER FROM MELTED ICE CUBE

5. The air-filled balloon is weighted so it sinks in water. Near the surface, the balloon has a certain volume. Draw the balloon at the bottom (inside the dashed square) and show whether it is bigger, smaller, or the same size.

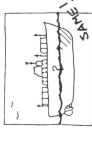

a. Since the weighted balloon sinks, how does its overall density compare to the density of water?

THE DENSITY OF BALLOON IS GREATER.

b. As the weighted balloon sinks, does its density increase, decrease, or remain the same?

DENSITY INCREASES (BEFORE VOLUME DECREASES)

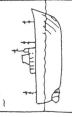

c. Since the weighted balloon sinks, how does the buoyant force on it compare to its weight?

BF IS LESS THAN ITS WEIGHT

d. As the weighted balloon sinks deeper, does the buoyant force on it increase, decrease, or remain the same?

BF DECREASES (BECAUSE VOLUME DECREASES)

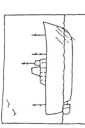

6. What would be your answers to Questions *a, b, c,* and *d* for a rock instead of an air-filled balloon?

a. DENSITY OF ROCK IS GREATER

b. DENSITY REMAINS SAME (SAME VOL)

c. BF IS LESS THAN ITS WEIGHT

d. BF STAYS SAME (VOL STAYS SAME)

CONCEPTUAL PHYSICAL SCIENCE EXPLORATIONS

Appendix D Fluid Physics

Gases

1. A principle difference between a liquid and a gas is that when a liquid is under pressure, its volume

(increases) (decreases) (**doesn't change noticeably**)

and its density

(increases) (decreases) (**doesn't change noticeably**)

When a gas is under pressure, its volume

(increases) (**decreases**) (doesn't change noticeably)

and its density

(**increases**) (decreases) (doesn't change noticeably)

2. The sketch shows the launching of a weather balloon at sea level. Make a sketch of the same weather balloon when it is high in the atmosphere. In words, what is different about its size and why?

Balloon grows as it rises. Atm pressure tends to compress things – even balloons. More pressure at ground level, + more compression at high altitudes, + bigger balloon.

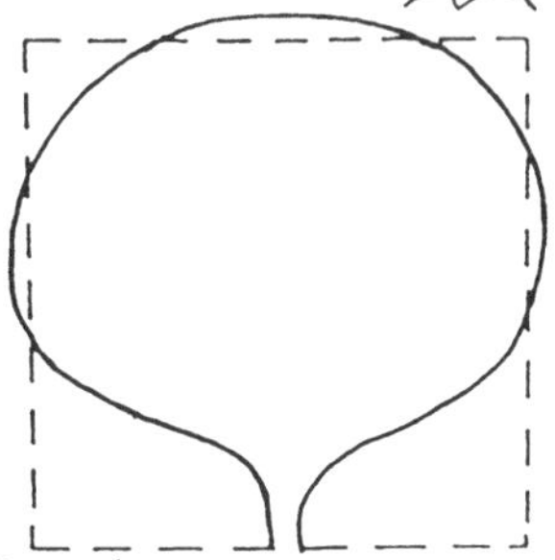

Exploring Conceptual Physical Science – *sigh*

↑ HIGH - ALTITUDE SIZE

GROUND - LEVEL SIZE

3. A hydrogen-filled balloon that weighs 10 N must displace __10__ N of air in order to float in air.

If it displaces less than __10__ N it will be buoyed up with less than __10__ N and sink

If it displaces more than __10__ N of air it will move upward.

4. Why is the cartoon more humorous to physics types than to non-physics types? What physics has occurred?

In accord with Bernoulli's Principle, movement of air over curved top of umbrella causes a reduction of air pressure (like airplane wing). This likely produced a net upward force that turned the umbrella inside out.

RATS TO YOU TOO, DANIEL BERNOULLI!

151

214